Alexander Bulovyatov

Efficient band structure computation of 3D photonic crystals

Alexander Bulovyatov

Efficient band structure computation of 3D photonic crystals

A parallel multigrid method with higher order modified finite elements for the 3D periodic eigenvalue problem in the curl-curl formulation

Südwestdeutscher Verlag für Hochschulschriften

Impressum/Imprint (nur für Deutschland/ only for Germany)
Bibliografische Information der Deutschen Nationalbibliothek: Die Deutsche Nationalbibliothek verzeichnet diese Publikation in der Deutschen Nationalbibliografie; detaillierte bibliografische Daten sind im Internet über http://dnb.d-nb.de abrufbar.

Verlag: Südwestdeutscher Verlag für Hochschulschriften Aktiengesellschaft & Co. KG
Dudweiler Landstr. 99, 66123 Saarbrücken, Deutschland
Telefon +49 681 37 20 271-1, Telefax +49 681 37 20 271-0
Email: info@svh-verlag.de
Zugl.: Karlsruhe, KIT, Diss., 2010

Herstellung in Deutschland:
Schaltungsdienst Lange o.H.G., Berlin
Books on Demand GmbH, Norderstedt
Reha GmbH, Saarbrücken
Amazon Distribution GmbH, Leipzig
ISBN: 978-3-8381-1840-6

Imprint (only for USA, GB)
Bibliographic information published by the Deutsche Nationalbibliothek: The Deutsche Nationalbibliothek lists this publication in the Deutsche Nationalbibliografie; detailed bibliographic data are available in the Internet at http://dnb.d-nb.de.

Publisher: Südwestdeutscher Verlag für Hochschulschriften Aktiengesellschaft & Co. KG
Dudweiler Landstr. 99, 66123 Saarbrücken, Germany
Phone +49 681 37 20 271-1, Fax +49 681 37 20 271-0
Email: info@svh-verlag.de

Printed in the U.S.A.
Printed in the U.K. by (see last page)
ISBN: 978-3-8381-1840-6

Contents

Abstract

This thesis considers the computation of the band structure (spectrum) of 3D photonic crystals. Photonic crystals are artificial periodic dielectric materials, which exhibit some special electromagnetic properties, e.g. waves of certain frequencies cannot propagate. A wide application of the photonic crystals in optics, lasers and electronics may push technologies to the next level.

Mathematically the problem turns into solving a family of Maxwell eigenvalue problems on the periodicity domain with an appropriate boundary conditions. The transformation from the whole space to a bounded periodic domain is done by the Bloch-Floquet theory and results in appearance of the shifted operators $\nabla + i\mathbf{k}$. The problem is considered in the frequency domain and formulated in a mixed form for the $\mathbf{H}$(curl) space, which is a natural space for 3D Maxwell equations.

For the discretization we use the finite element method with special $\mathbf{H}$(curl)- and H^1-conforming k-modified elements. They allow to adapt the usual $\mathbf{H}$(curl)- and H^1-conforming discretizations for the case of the shifted operators $\nabla + i\mathbf{k}$ and the periodic boundary condition. It is required in order to obtain a correct numerical approximation of the spectrum. We use the lowest and higher order hierarchical finite element construction with the local exact sequence property, which provides a useful representation of gradient fields at the level of shape functions.

The eigenvalue problem is solved by a preconditioned iterative eigenvalue solver, which is a modification of the LOBPCG method extended by a projection onto the divergence-free vector fields, what allows to apply the solver to the Maxwell curl curl operator with a large kernel. The projection is based on the discrete Helmholtz decomposition and is performed in the potential space by solving an auxiliary linear Laplace problem with a multigrid iterative solver. The preconditioning of the eigenvalue solver is done by solving a regularized linear Maxwell problem with a multigrid iterative solver.

The multigrid method for the regularized Maxwell problem is based on a multilevel nodal decomposition, which results in the standard multigrid V-cycle with a special hybrid smoother. The hybrid smoother respects the discrete Helmholtz decomposition and provides robust convergence rate of the Maxwell multigrid method. In the case of higher order finite elements the Maxwell multigrid and the Laplace multigrid are realized as a two-level p-multigrid method with the lowest order elements at the coarse level.

Band structure computations of the 3D photonic crystal require a lot of computing power, the multigrid solvers are the most computation-intensive parts. We use a parallel multigrid method for the distributed memory model (MPI parallelization) based on a concept of the geometric-centered data structures. This approach leads to an efficient and highly scalable software implementation, which suits well for modern high performance parallel clusters.

The theory is confirmed by the numerical computations. We provide results of the eigenvalue computations for two different 3D crystal structures with a full band gap, obtained with the lowest and higher order finite elements. We compute the band structures and perform the computer-assisted perturbation proof for band gap. The eigenvalue solver convergence, the convergence of the finite elements and the parallel performance are discussed.

The main contribution of this work is a complete description of the theory, algorithms and implementation issues required for band structure computations of 3D photonic crystals. Many results for the standard Maxwell formulation have to be adopted for the periodic formulation. It leads to significant changes in the finite element construction (k-modified elements), what results in the changes in representation of differential operators and the multigrid interpolation / restriction operations. As far as we know, these issues and the parallel implementation have not been described in detail before. We also present an improved LOBPCG algorithm and the perturbation proof for spectrum of the 3D Maxwell operator, which is a part of the computer-assisted proof for band gap. All algorithms in the thesis are realized as a part of the free open source parallel finite element software "M++".

Acknowledgements

I devote the thesis to my mother Galina. She has been believing in me since I made my first step. In the difficult 90s she did her best for my upbringing and education.

I am very grateful to Deutsche Forschungsgemeinschaft for the PhD scholarship. This financial support made it possible to write the thesis.

Special thank goes to my scientific advisor Prof. Dr. Christian Wieners for the competent supervising and inspiring my work. He was always ready to spend some time discussing current problems. I believe I have learnt a lot from him.

It was pleasure for me to work in GRK 1294 and IWRMM among many bright people. I thank Dr. Wolfgang Müller and Mr. Daniel Maurer for the collaboration and co-development of the software. I am appreciated to Prof. Dr. Willy Dörfler, Dr. Kai Sandfort and Dr. Arne Schneck for multiple scientific discussions and useful hints.

At last, but not least I thank my wife Tatiana for proof-reading the text and supporting me day-by-day.

Chapter 1

Introduction

1.1 Problem statement

1.1.1 Maxwell's equations

We start with the classical *macroscopic* Maxwell's equations in $\mathbb{R}^3$ written in SI units

$$\begin{aligned} \nabla \times \mathcal{E} &= -\frac{\partial \mathcal{B}}{\partial t}, \\ \nabla \cdot \mathcal{B} &= 0, \\ \nabla \times \mathcal{H} &= \frac{\partial \mathcal{D}}{\partial t} + \mathcal{J}, \\ \nabla \cdot \mathcal{D} &= \rho, \end{aligned} \tag{1.1}$$

where $\mathcal{E}$ is the *electric field intensity*, $\mathcal{B}$ is the *magnetic field intensity*, $\mathcal{D}$ is the *electric displacement field*, $\mathcal{H}$ is the *magnetic induction field*, ρ is the *charge density*, $\mathcal{J}$ is the *current density field*. All quantities are functions of position $\mathbf{x} \in \mathbb{R}^3$ and time $t \in \mathbb{R}$. The electromagnetic field is defined by four vector fields $\mathcal{E}$, $\mathcal{B}$, $\mathcal{D}$ and $\mathcal{H}$. The sources of the electromagnetic field are scalar field ρ and vector field $\mathcal{J}$. Henceforth a bold font means a vector quantity.

Maxwell's equations in electrodynamics represent a large scientific area, we will mention only the key facts and results. For physical explanation and detailed problem statement one may refer to a general book about electrodynamics, e.g. [20].

Usually photonic crystals are constructed of *dielectric* materials, for that special case we make some simplifications, namely, put $\rho = 0$ (no *free charge* is present) and $\mathcal{J} = 0$ (material is *non-conducting*).

We consider only the case of *linear* medium. For such a medium there are two linear *constitutive laws*, which relate $\mathcal{D}$ to $\mathcal{E}$ and $\mathcal{B}$ to $\mathcal{H}$ as

$$\begin{aligned} \mathcal{D} &= \varepsilon \mathcal{E}, \\ \mathcal{B} &= \mu \mathcal{H}. \end{aligned} \tag{1.2}$$

So, any linear dielectric material is determined by two properties: the *electric permittivity* ε and the *magnetic permeability* μ.

We assume that the medium has no frequency dependence (*material dispersion*), what means that ε and μ depend on position $\mathbf{x}$ only. Moreover, we suppose that the medium is *lossless*, i.e. no absorption of electromagnetic energy occurs inside the medium. It follows that for *isotropic* medium ε and μ are scalar real-valued positive functions. In a more general case of *anisotropic* medium they are matrix functions, for any $\mathbf{x} \in \mathbb{R}^3$ ε and μ are real positive definite 3×3 matrices. If the medium were lossy then ε and μ could have an imaginary part.

Since most of materials used for production of photonic crystals have no magnetic properties, we assume that $\mu = 1$. Laws (1.2) together with equations (1.1) let us obtain a simplified form, where we keep only two vector fields of four:

$$\begin{aligned} \nabla \times \mathcal{E} &= -\frac{\partial \mathcal{H}}{\partial t}, \\ \nabla \cdot \mathcal{H} &= 0, \\ \nabla \times \mathcal{H} &= \varepsilon \frac{\partial \mathcal{E}}{\partial t}, \\ \nabla \cdot (\varepsilon \mathcal{E}) &= 0. \end{aligned} \tag{1.3}$$

From the time-dependent problem we go to the *time-harmonic* form. In this case it is supposed that electromagnetic field is *monochromatic*, i.e. has one temporal frequency $\omega > 0$ and varies sine-like in time. This assumption simplifies the problem, but does not restrict generality, because from Fourier analysis it follows that any solution can be represented by harmonic modes. We apply the following ansatz:

$$\begin{aligned} \mathcal{E}(\mathbf{x}, \mathbf{t}) &= \mathrm{Re}\left(\mathrm{e}^{\mathrm{i}\omega t}\mathbf{E}(\mathbf{x})\right), \\ \mathcal{H}(\mathbf{x}, \mathbf{t}) &= \mathrm{Re}\left(\mathrm{e}^{\mathrm{i}\omega t}\mathbf{H}(\mathbf{x})\right), \end{aligned}$$

where $\mathbf{E}$ and $\mathbf{H}$ are complex vector fields. Combining the ansatz and equations (1.3) one gets a time-harmonic form of Maxwell's equations, where time t is excluded:

$$\nabla \times \mathbf{E} = -\mathrm{i}\omega \mathbf{H}, \tag{1.4}$$

$$\nabla \cdot \mathbf{H} = 0, \tag{1.5}$$

$$\nabla \times \mathbf{H} = \mathrm{i}\omega\varepsilon \mathbf{E}, \tag{1.6}$$

$$\nabla \cdot (\varepsilon \mathbf{E}) = 0. \tag{1.7}$$

The current system can be simplified further. It is possible to decouple the equations and extract a problem with only one vector field, let it be $\mathbf{H}$. From equation (1.6) one gets

$$\mathbf{E} = -\frac{\mathrm{i}}{\omega}\varepsilon^{-1}\nabla \times \mathbf{H}, \tag{1.8}$$

together with equations (1.4) and (1.5) it gives

$$\nabla \times \left(\varepsilon^{-1}\nabla \times \mathbf{H}\right) = \omega^2 \mathbf{H}, \tag{1.9}$$

$$\nabla \cdot \mathbf{H} = 0. \tag{1.10}$$

The second equation can be derived from the first one for $\omega > 0$, but we keep it for the moment since this property is a very important one. We consider equation (1.9) as an *eigenvalue problem*, so for a given ε we look for eigenvalues $\lambda = \omega^2$ and eigenfunctions $\mathbf{H}$. As soon as this problem is solved and $\mathbf{H}$ is found, $\mathbf{E}$ may be restored by formula (1.8).

1.1.2 Periodicity, Bloch-Floquet theory

Photonic crystals are a kind of *periodic structures*. In modelling we assume that a crystal is unbounded and occupies the whole $\mathbb{R}^3$ space. It is an abstraction developed from the idea that the whole crystal is large with respect to one element of the periodic structure. Different physical aspects of crystal structures are studied in the book [22], a useful overview of mathematical approaches is given in [27].

In our model, a photonic crystal is completely defined by distribution of the electric permittivity ε, which is uniformly bounded away from zero and has certain periodicity. Let $d \in \{1, 2, 3\}$, suppose that there exist linearly independent vectors $\mathbf{r}_1, \dots, \mathbf{r}_d \in \mathbb{R}^3$ s.t.

$$\varepsilon(\mathbf{x}) = \varepsilon(\mathbf{x} + \mathbf{r}_j) \qquad \text{for all } \mathbf{x} \in \mathbb{R}^3 \quad j = 1, \dots, d, \tag{1.11}$$

then the medium is called d-dimensional periodic medium (photonic crystal), $\{\mathbf{r}_j\}$ with the minimal lengths are called the *primitive vectors.* For such a medium one may define d-dimensional *Bravais lattice*

$$\Lambda = \left\{ \sum_{j=1}^{d} l_j \mathbf{r}_j \mid l_1, \dots, l_d \in \mathbb{Z} \right\}.$$

(1.11) means that the medium actually has a certain unique domain and all space consists of copies of that domain. A d-dimensional domain Ω is called *fundamental domain* if for any $\mathbf{x} \in \mathbb{R}^d$ there exists $\mathbf{a} \in \Lambda$ s.t. either $\mathbf{a}$ is unique and $\mathbf{x} + \mathbf{a} \in \Omega$, or $\mathbf{a}$ is not unique and $\mathbf{x} + \mathbf{a} \in \partial\Omega$. So we can write

$$\mathbb{R}^d = \bigcup_{\mathbf{a} \in \Lambda} (\overline{\Omega} + \mathbf{a}),$$

where $\overline{\Omega} + \mathbf{a}_1$ and $\overline{\Omega} + \mathbf{a}_2$ may have intersection only along their boundaries. In solid-state physics such a domain is called *Wigner-Seitz cell.* An example of a fundamental domain is a domain that consists of $\mathbf{x} \in \mathbb{R}^d$ which are closer to the origin than to any $\mathbf{a} \in \Lambda$.

Shape of the fundamental domain and distribution of materials inside the domain may have *symmetries,* e.g. ones with respect to rotations and reflections. It is an important aspect, we will see later that the symmetries can significantly reduce amount of computations needed to solve the Maxwell eigenvalue problem.

From mathematical point of view problem (1.9) is a *system of partial differential equations* with periodic coefficients. The main tool to work with such a kind of equations is *Floquet-Bloch theory.* For a detailed explanation about this theory we refer to [26], one of the early publications with proofs is [33].

For a set of primitive vectors $\{\mathbf{r}_j\}$ let us define linearly independent vectors $\hat{\mathbf{r}}_1, \dots, \hat{\mathbf{r}}_d \in \mathbb{R}^3$ s.t.

$$\mathbf{r}_i \cdot \hat{\mathbf{r}}_j = 2\pi\delta_{ij} \qquad \text{for any } i, j \in \{1, \dots, d\},$$

they are called *primitive reciprocal vectors.* The d-dimensional *reciprocal lattice* $\hat{\Lambda}$ is defined by

$$\hat{\Lambda} = \left\{ \sum_{j=1}^{d} l_j \hat{\mathbf{r}}_j \mid l_1, \dots, l_d \in \mathbb{Z} \right\}.$$

By analogy with Wigner-Seitz cell, the domain K consisted of $\mathbf{k} \in \mathbb{R}^d$ which are closer to the origin than to any $\hat{\mathbf{a}} \in \hat{\Lambda}$ is called the (first) *Brillouin zone.*

We consider a function $f\colon \mathbb{R}^d \to \mathbb{C}$ which is defined on a periodic medium. Provided that $|f(\mathbf{x})|$ decays sufficiently fast while $\|\mathbf{x}\| \to \infty$, we define the *Floquet transform* $\mathcal{U}_\Lambda f$ of function f with respect to the lattice Λ by formula

$$(\mathcal{U}_\Lambda f)(\mathbf{x}, \mathbf{k}) = \frac{1}{\sqrt{|K|}} \sum_{\mathbf{a} \in \Lambda} f(\mathbf{x} - \mathbf{a}) \mathrm{e}^{\mathrm{i}\mathbf{k}\cdot\mathbf{a}}, \tag{1.12}$$

where $\mathbf{k} \in \mathbb{R}^d$ is called *quasi-momentum.*

One can easily notice the following properties of the Floquet transform. For any $\mathbf{x}, \mathbf{k} \in \mathbb{R}^d, \mathbf{a} \in \Lambda, \hat{\mathbf{a}} \in \hat{\Lambda}$

$$(\mathcal{U}_\Lambda f)(\mathbf{x} + \mathbf{a}, \mathbf{k}) = \mathrm{e}^{\mathrm{i}\mathbf{k}\cdot\mathbf{a}} (\mathcal{U}_\Lambda f)(\mathbf{x}, \mathbf{k}), \tag{1.13}$$

$$(\mathcal{U}_\Lambda f)(\mathbf{x}, \mathbf{k} + \hat{\mathbf{a}}) = (\mathcal{U}_\Lambda f)(\mathbf{x}, \mathbf{k}). \tag{1.14}$$

The first relation is the *Floquet condition*, it means that it is sufficient to know $(\mathcal{U}_\Lambda f)(\cdot, \mathbf{k})$ on the closure of a fundamental domain Ω to extend it to the whole $\mathbb{R}^d$. The second relation exposes that the transform is periodic with respect to quasi momentum $\mathbf{k}$, so it is enough to know $(\mathcal{U}_\Lambda f)(\mathbf{x}, \cdot)$ on the closure of the first Brillouin zone K. Therefore we will consider that the Floquet transform is defined only for $\mathbf{x} \in \overline{\Omega}$ and $\mathbf{k} \in \overline{K}$.

Let $\mathcal{L}(\mathbf{x}, \nabla_\mathbf{x})$ be a *linear uniformly elliptic partial differential operator* with Λ-periodic coefficients. The operator is associated with $L(\mathbf{x}, \nabla_\mathbf{x})$, a *symbolic form of differential operator*, which has no domain. Due to periodicity the operator commutes with the transform

$$\mathcal{U}_\Lambda(L(\mathbf{x}, \nabla_\mathbf{x})f) = L(\mathbf{x}, \nabla_\mathbf{x})(\mathcal{U}_\Lambda f). \tag{1.15}$$

Let us discuss the right hand side of the relation. The differential operator is applied to a function of $\mathbf{x}$ and $\mathbf{k}$, but the derivative is taken with respect to $\mathbf{x}$ only. Moreover, $(\mathcal{U}_\Lambda f)(\cdot, \mathbf{k})$ is a function which holds the Floquet condition (1.13) on $\partial\Omega$. Although the differential expression of the operator is independent of $\mathbf{k}$, its domain is different for every $\mathbf{k}$. So, we conclude that the Floquet transform turns the periodic differential operator $\mathcal{L}(\mathbf{x}, \nabla_\mathbf{x})$ acting on $\{f : \mathbb{R}^d \to \mathbb{C}\}$ to a family of differential operators $\mathcal{L}_\mathbf{k} = L(\mathbf{x}, \nabla_\mathbf{x})$ acting on

$$D(\mathcal{L}_\mathbf{k}) = \{g : \overline{\Omega} \to \mathbb{C} \mid g(\mathbf{x} + \mathbf{a}) = e^{i\mathbf{k}\cdot\mathbf{a}} g(\mathbf{x}),\ \mathbf{a} \in \Lambda,\ \mathbf{x}, \mathbf{x} + \mathbf{a} \in \partial\Omega\}.$$

It is important that now the functions are defined on a *compact* manifold $\overline{\Omega}$.

Let us introduce a complex *Hilbert space* $L^2(\mathbb{R}^d)$ with the *inner product*

$$\langle u, v\rangle = \int_{\mathbb{R}^d} u(\mathbf{x})\overline{v(\mathbf{x})}\, d\mathbf{x}, \qquad u, v \in L^2(\mathbb{R}^d).$$

Theorem 1.1. *by Paley-Wiener (see [27, Theorem 7.2])*
The transform $\mathcal{U}_\Lambda : L^2(\mathbb{R}^d) \to L^2(K, L^2(\overline{\Omega}))$ is an isometric isomorphism. Its inverse transform is defined by

$$(\mathcal{U}_\Lambda^{-1} g)(\mathbf{x}) = \frac{1}{\sqrt{|K|}} \int_K g(\mathbf{x}, \mathbf{k}) dk,$$

where $g(\cdot, \mathbf{k}) \in D(\mathcal{L}_\mathbf{k})$ for a fixed $\mathbf{k} \in K$.

We assume that the operator $L(\mathbf{x}, \nabla_\mathbf{x})$ is *formally self-adjoint* with respect to $\langle\cdot,\cdot\rangle$. For a fixed $\mathbf{k} \in \overline{K}$ we consider an eigenvalue problem for the operator $\mathcal{L}_\mathbf{k}$

$$\begin{aligned} L(\mathbf{x}, \nabla_\mathbf{x})\psi &= \lambda\psi, \\ \psi(\mathbf{x} + \mathbf{a}) &= e^{i\mathbf{k}\cdot\mathbf{a}}\psi(\mathbf{x}), \qquad \mathbf{a} \in \Lambda,\ \mathbf{x}, \mathbf{x} + \mathbf{a} \in \partial\Omega. \end{aligned} \tag{1.16}$$

It is a symmetric eigenvalue problem in $L^2(\Omega)$, where $\overline{\Omega}$ is a compact manifold. Together with the *ellipticity condition* for $L(\mathbf{x}, \nabla_\mathbf{x})$ it follows that the operators $\mathcal{L}_\mathbf{k}$ have compact resolvents and hence discrete spectrum

$$0 < \lambda_1(\mathbf{k}) \le \ldots \le \lambda_s(\mathbf{k}) \to \infty \quad \text{as } s \to \infty,$$

with orthonormal and complete set of eigenfunctions $\psi_1(\cdot; \mathbf{k}), \ldots, \psi_s(\cdot; \mathbf{k})$, which are called *Bloch modes*.

Using the Floquet transform and Theorem 1.1 one can prove completeness of the set of the Bloch modes. The next theorem formulates it as follows.

Theorem 1.2. *(see [33, Theorem 1])*
The set of the Bloch modes $\{\psi_j(\mathbf{x}; \mathbf{k})\}$, where $\mathbf{k}$ varies over the Brillouin zone K and $j \in \mathbb{N}$, is complete.

For any $f \in L^2(\mathbb{R}^d)$ and $l \in \mathbb{N}$, define

$$f_l(\mathbf{x}) = \frac{1}{\sqrt{|K|}} \sum_{s=1}^{l} \int_K \langle (\mathcal{U}_\Lambda f)(\mathbf{x}, \mathbf{k}), \psi_s(\mathbf{x}; \mathbf{k}) \rangle \psi_s(\mathbf{x}; \mathbf{k}) dk.$$

Then $f_l \to f$ in $L^2(\mathbb{R}^d)$ while $l \to \infty$.

We may define $\phi_s(\mathbf{x}; \mathbf{k}) = \mathrm{e}^{-\mathrm{i}\mathbf{k}\cdot\mathbf{x}} \psi_s(\mathbf{x}; \mathbf{k})$. One can check that

$$\nabla_{\mathbf{x}} \psi_s(\mathbf{x}; \mathbf{k}) = \mathrm{e}^{\mathrm{i}\mathbf{k}\cdot\mathbf{x}} (\nabla_{\mathbf{x}} + \mathrm{i}\mathbf{k}) \phi_s(\mathbf{x}; \mathbf{k}).$$

It allows us to rewrite (1.16) in the form

$$\begin{aligned} L(\mathbf{x}, \nabla_{\mathbf{x}} + \mathrm{i}\mathbf{k})\phi &= \lambda\phi, \\ \phi(\mathbf{x} + \mathbf{a}) &= \phi(\mathbf{x}), \qquad \mathbf{a} \in \Lambda,\ \mathbf{x}, \mathbf{x} + \mathbf{a} \in \partial\Omega. \end{aligned} \tag{1.17}$$

So, from the fixed differential operator $\mathcal{L}_{\mathbf{k}}$ with a k-dependent domain (the quasi-periodic boundary condition) we go to the k-dependent differential operator $\mathcal{L}_{\mathbf{k}}^{\mathrm{per}} = L(\mathbf{x}, \nabla_{\mathbf{x}} + \mathrm{i}\mathbf{k})$ with a fixed domain (the periodic boundary condition). The eigenvalues stay the same, the completeness property remains.

1.1.3 Spectrum of operator, band structure

According to (1.15) and Theorem 1.1, the Floquet transform expands the self-adjoint periodic operator $\mathcal{L}(\mathbf{x}, \nabla_{\mathbf{x}})$ into the direct integral of the operators $\mathcal{L}_{\mathbf{k}}$. One can prove the following representation of the spectrum (see [26])

$$\sigma\big(\mathcal{L}(\mathbf{x}, \nabla_{\mathbf{x}})\big) = \bigcup_{\mathbf{k} \in \overline{K}} \sigma\big(\mathcal{L}_{\mathbf{k}}^{\mathrm{per}}(\mathbf{x}, \nabla_{\mathbf{x}} + \mathrm{i}\mathbf{k})\big) = \overline{\bigcup_{\mathbf{k} \in S} \sigma\big(\mathcal{L}_{\mathbf{k}}^{\mathrm{per}}(\mathbf{x}, \nabla_{\mathbf{x}} + \mathrm{i}\mathbf{k})\big)},$$

where S is a dense subset in K. The latter representation allows us to drop a finite set of $\mathbf{k}$ points (e.g. $\mathbf{k} = \mathbf{0}$) from $\overline{K}$ without affecting the spectrum.

For each $j \in \mathbb{N}$ an eigenvalue $\lambda_j(\mathbf{k})$ of the operator $\mathcal{L}_{\mathbf{k}}^{\mathrm{per}}$ is a continuous function of parameter $\mathbf{k} \in K$ (see [26]). This follows from the fact that the coefficients in the problem (1.17) depend continuously on $\mathbf{k}$. $\lambda_j(\cdot)$ is called *a band function* and its graph is called *a band.* For $j \in \mathbb{N}$ define $I_j = \{\lambda_j(\mathbf{k}) \mid \mathbf{k} \in \overline{K}\}$, since $\overline{K}$ is compact and connected, I_j is a compact real interval. It gives another representation of the spectrum

$$\sigma\big(\mathcal{L}(\mathbf{x}, \nabla_{\mathbf{x}})\big) = \bigcup_{j \in \mathbb{N}} I_j.$$

If there exist some $m \in \mathbb{N}$ and an interval (a, b) s.t. for any $x \in I_m, y \in (a, b), z \in I_{m+1}$ we have $x < y < z$, then the interval (a, b) is called *a band gap* between the bands m and $m + 1$. In two and especially in three dimensions different I_j usually are significantly overlapping. In order "to open" a band gap one needs a special material distribution with high contrast in ε (see [27]).

A very important practical aspect is redundancy of the bands due to symmetries. Let $G\colon \mathbb{R}^3 \to \mathbb{R}^3$ be an orthonormal linear operator (in a basis it can be represented as a rotation with respect to some axis and / or reflections with respect to some planes) s.t. $\varepsilon(G\mathbf{x}) = \varepsilon(\mathbf{x})$ for all $\mathbf{x} \in \overline{\Omega}$. One may check that if (ϕ, λ) is a solution of the problem (1.17) for some $\mathbf{k} = \mathbf{k}_0$, then $(\phi \circ G^{-1}), \lambda)$ is a solution of the problem for $\mathbf{k} = G\,\mathbf{k}_0$ (see e.g. [21, Chapter 3]).

We conclude that if there are symmetries in the material distribution ε, then we have some redundancy within the Brillouin zone. In order to get the full spectrum we do not need to

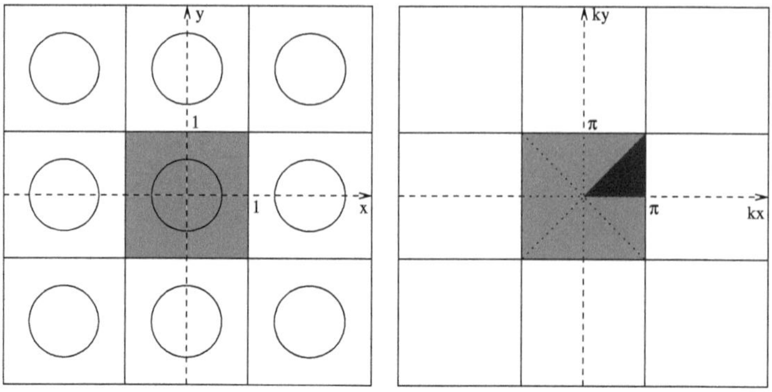

Figure 1.1: A fundamental domain (left) and the corresponding Brillouin zone (right) with the triangular irreducible Brillouin zone.

consider the problem (1.17) for all $\mathbf{k} \in \overline{K}$. The smallest region within the Brillouin zone which cannot be obtained via symmetries is called the *irreducible Brillouin zone*. For example, a simple square lattice made of circular rods has a square Brillouin zone, but the irreducible Brillouin zone is just a triangle with the area only 1/8 of the whole zone, since the lattice is symmetric with respect to three reflections: $x \to -x$, $y \to -y$ and $x \to y$, so the same reflections cut the Brillouin zone. The situation is shown at Figure 1.1.

1.1.4 Maxwell equations in periodic media

We assume that ε is a positive piecewise continuous scalar function. For simplicity let us take the lattice $\Lambda = \mathbb{Z}^3$, so

$$\varepsilon(\mathbf{x}) = \varepsilon(\mathbf{x}+\mathbf{z}), \qquad \text{for all } \mathbf{z} \in \Lambda, \quad \mathbf{x} \in \mathbb{R}^3.$$

Let the fundamental domain be $\Omega = [0,1]^3$, so the corresponding first Brillouin zone is $K = [-\pi,\pi]^3$.

We make the ansatz in form of the Bloch modes

$$\mathbf{H}(\mathbf{x}) = e^{i\mathbf{k}\cdot\mathbf{x}}\widetilde{\mathbf{H}}(\mathbf{x}), \tag{1.18}$$

where $\mathbf{k} \in K$ and $\widetilde{\mathbf{H}}(\mathbf{x})$ is a periodic function in Ω, i.e. for all $x_1, x_2, x_3 \in [0,1]$

$$\begin{aligned}
\widetilde{\mathbf{H}}(x_1,x_2,1) &= \widetilde{\mathbf{H}}(x_1,x_2,0),\\
\widetilde{\mathbf{H}}(x_1,1,x_3) &= \widetilde{\mathbf{H}}(x_1,0,x_3),\\
\widetilde{\mathbf{H}}(1,x_2,x_3) &= \widetilde{\mathbf{H}}(0,x_2,x_3).
\end{aligned}$$

Inserting the ansatz (1.18) into the Maxwell equations (1.9) we obtain the following formulation (assume ε is smooth for the moment).

$$\begin{aligned}
\nabla_{\mathbf{k}} \times \left(\varepsilon^{-1}\, \nabla_{\mathbf{k}} \times \widetilde{\mathbf{H}}\right) &= \omega^2 \widetilde{\mathbf{H}} \qquad \text{in } \Omega,\\
\nabla_{\mathbf{k}} \cdot \widetilde{\mathbf{H}} &= \mathbf{0} \qquad \text{in } \Omega,
\end{aligned}$$

where $\nabla_{\mathbf{k}} = \nabla + i\mathbf{k}$. Later we use $\omega^2 = \lambda$ to simplify the notations.

The formulation in form of a partial differential equations does not allow to handle the practical important case of a jumping ε. An integral formulation is more appropriate for this. We need to introduce the periodic versions of the usual Sobolev spaces. Let $L^2(\Omega) = L^2(\Omega, \mathbb{C})$ and $\mathbf{L}^2(\Omega) = L^2(\Omega, \mathbb{C}^3)$ be scalar and vector valued L^2-functions.

Definition 1. Let Ω be a *bounded, simply-connected Lipschitz-domain.* Define spaces of infinitely differentiable functions

$$
\begin{aligned}
C^\infty(\overline{\Omega}) &= \{f : \overline{\Omega} \to \mathbb{C} \mid \text{there exists } \frac{\partial^{l+m+n} f}{\partial x^l \partial y^m \partial z^n} \text{ for any } l, m, n \in \{0\} \cup \mathbb{N}\},\\
C_1^\infty(\overline{\Omega}) &= \{f \in C^\infty(\overline{\Omega}) \mid f(\mathbf{x}+\mathbf{a}) = f(\mathbf{x}), \quad \mathbf{a} \in \Lambda;\ \mathbf{x}, \mathbf{x}+\mathbf{a} \in \partial\Omega\},\\
C_{\text{curl}}^\infty(\overline{\Omega}) &= \{\mathbf{u} \in (C^\infty(\overline{\Omega}))^3 \mid \mathbf{u}(\mathbf{x}+\mathbf{a}) \times \mathbf{n} = -\mathbf{u}(\mathbf{x}) \times \mathbf{n},\ \mathbf{a} \in \Lambda;\ \mathbf{x}, \mathbf{x}+\mathbf{a} \in \partial\Omega\},\\
C_{\text{div}}^\infty(\overline{\Omega}) &= \{\mathbf{u} \in (C^\infty(\overline{\Omega}))^3 \mid \mathbf{u}(\mathbf{x}+\mathbf{a}) \cdot \mathbf{n} = -\mathbf{u}(\mathbf{x}) \cdot \mathbf{n},\ \mathbf{a} \in \Lambda;\ \mathbf{x}, \mathbf{x}+\mathbf{a} \in \partial\Omega\},
\end{aligned}
$$

where $\mathbf{n}$ is the normal vector at the boundary. Define inner products

$$
\begin{aligned}
(f, g)_{L^2} &= \int_\Omega f\overline{g}\, d\mathbf{x},\\
(f, g)_{H^1} &= \int_\Omega \nabla f \cdot \overline{\nabla g}\, d\mathbf{x} + (f, g)_{L^2},\\
(\mathbf{u}, \mathbf{v})_{\mathbf{H}(\text{curl})} &= \int_\Omega \nabla \times \mathbf{u} \cdot \overline{\nabla \times \mathbf{v}}\, d\mathbf{x} + (\mathbf{u}, \mathbf{v})_{\mathbf{L}^2},\\
(\mathbf{u}, \mathbf{v})_{\mathbf{H}(\text{div})} &= \int_\Omega (\nabla \cdot \mathbf{u})\overline{(\nabla \cdot \mathbf{v})}\, d\mathbf{x} + (\mathbf{u}, \mathbf{v})_{\mathbf{L}^2},
\end{aligned}
$$

they induce the norms $\|\cdot\|_{H^1}$, $\|\cdot\|_{\mathbf{H}(\text{curl})}$, $\|\cdot\|_{\mathbf{H}(\text{div})}$. Now construct periodic spaces as closures with respect to the corresponding norms

$$
\begin{aligned}
H^1_{\text{per}}(\Omega) &= \overline{C_1^\infty(\overline{\Omega})}^{\|\cdot\|_{H^1}},\\
\mathbf{H}_{\text{per}}(\text{curl}, \Omega) &= \overline{C_{\text{curl}}^\infty(\overline{\Omega})}^{\|\cdot\|_{\mathbf{H}(\text{curl})}},\\
\mathbf{H}_{\text{per}}(\text{div}, \Omega) &= \overline{C_{\text{div}}^\infty(\overline{\Omega})}^{\mathbf{H}(\text{div})}.
\end{aligned}
$$

1.1.5 The main properties of $\nabla_\mathbf{k}$

In this thesis we work with the modified operator $\nabla_\mathbf{k}$. Due to the periodicity of medium we have to use it all the time, so it is important to summarize some of its properties. In [13] it is shown that using the operator $\nabla_\mathbf{k}$ we still have the most of the properties associated with the standard operator ∇.

Note that we always assume $\mathbf{k} \neq \mathbf{0}$. When $\mathbf{k} = \mathbf{0}$ it is a special regime, which leads to some complications. In the computations we avoid that by approximating $\mathbf{k} = \mathbf{0}$ with a vector of very small magnitude.

Theorem 1.3. *Helmholtz decomposition. (see [13, Theorem 3.1])*
Let $\mathbf{k} \neq \mathbf{0}$, for any given $\mathbf{u} \in \mathbf{L}^2(\Omega)$, there exist unique functions $\mathbf{v} \in \mathbf{H}^1_{\text{per}}(\Omega)$ and $q \in H^1_{\text{per}}(\Omega)$ s.t.

$$
\begin{aligned}
\mathbf{u} - \nabla_\mathbf{k} \times \mathbf{v} + \nabla_\mathbf{k}\, q,\\
\nabla_\mathbf{k} \cdot \mathbf{v} = 0.
\end{aligned}
$$

Moreover, for some $s \geq 0$ it holds

$$
\begin{aligned}
\|\mathbf{v}\|_1 + \|q\|_1 &\leq C\|\mathbf{u}\|,\\
\|\mathbf{v}\|_{1+s} &\leq C\| \nabla_\mathbf{k} \times \mathbf{v}\|_s,\\
\|q\|_{1+s} &\leq C\| \nabla_\mathbf{k}\, q\|_s.
\end{aligned}
$$

Corollary 1.1. From the theorem it easily follows that

$$\begin{aligned} q = 0 \quad &\Leftrightarrow \quad \nabla_{\mathbf{k}} \cdot \mathbf{u} = 0, \\ \mathbf{v} = \mathbf{0} \quad &\Leftrightarrow \quad \nabla_{\mathbf{k}} \times \mathbf{u} = \mathbf{0}. \end{aligned}$$

Corollary 1.2. (see [13]) The following sequence is exact, i.e. the range of each differential operator coincides with the kernel of the following operator.

$$H^1_{\text{per}}(\Omega) \xrightarrow{\nabla_{\mathbf{k}}} \mathbf{H}_{\text{per}}(\text{curl}, \Omega) \xrightarrow{\nabla_{\mathbf{k}} \times} \mathbf{H}_{\text{per}}(\text{div}, \Omega) \xrightarrow{\nabla_{\mathbf{k}} \cdot} L^2(\Omega).$$

1.2 The eigenvalue problem

For $\mathbf{u}, \mathbf{v} \in \mathbf{H}_{\text{per}}(\text{curl}, \Omega)$ and $p, q \in H^1_{\text{per}}(\Omega)$ we define the sesquilinear forms

$$\begin{aligned} m(\mathbf{u}, \mathbf{v}) &= (\mathbf{u}, \mathbf{v})_{\mathbf{L}^2} = \int_\Omega \mathbf{u} \cdot \overline{\mathbf{v}}\, d\mathbf{x} \\ a_{\mathbf{k}}(\mathbf{u}, \mathbf{v}) &= (\varepsilon^{-1} \nabla_{\mathbf{k}} \times \mathbf{u}, \nabla_{\mathbf{k}} \times \mathbf{v})_{\mathbf{L}^2}, \\ b_{\mathbf{k}}(\mathbf{v}, q) &= (\mathbf{v}, \nabla_{\mathbf{k}}\, q)_{\mathbf{L}^2}, \\ c_{\mathbf{k}}(p, q) &= (\nabla_{\mathbf{k}}\, p, \nabla_{\mathbf{k}}\, q)_{\mathbf{L}^2}. \end{aligned}$$

We define the constraint space

$$\mathbf{V}_{\mathbf{k}} = \{\mathbf{v} \in \mathbf{H}_{\text{per}}(\text{curl}, \Omega) \colon b_{\mathbf{k}}(\mathbf{v}, q) = 0 \quad \text{for all } q \in H^1_{\text{per}}(\Omega)\},$$

so $\mathbf{V}_{\mathbf{k}} = \mathbf{H}_{\text{per}}(\text{curl}, \Omega) \cap (\nabla_{\mathbf{k}} H^1_{\text{per}}(\Omega))^\perp$. The forms $m(\cdot,\cdot)$, $a_{\mathbf{k}}(\cdot,\cdot)$, $c_{\mathbf{k}}(\cdot,\cdot)$ are Hermitian, $m(\cdot,\cdot)$ is positive definite.

Define operator $A_{\mathbf{k}} \colon \mathbf{H}_{\text{per}}(\text{curl}, \Omega) \to \mathbf{L}^2(\Omega)$ s.t. $\langle A_{\mathbf{k}}\mathbf{u}, \mathbf{v}\rangle = a_{\mathbf{k}}(\mathbf{u}, \mathbf{v})$ for all $\mathbf{v} \in \mathbf{H}_{\text{per}}(\text{curl}, \Omega)$. From Corollary 1.2 we have that $\ker(A_{\mathbf{k}}) = \ker(\nabla_{\mathbf{k}} \times) = \nabla_{\mathbf{k}} H^1_{\text{per}}(\Omega)$, so the operator $A_{\mathbf{k}}$ is positive definite on $\mathbf{V}_{\mathbf{k}}$ if $\mathbf{k} \neq \mathbf{0}$.

Let us prove that $c_{\mathbf{k}}(\cdot,\cdot)$ is coercive for $\mathbf{k} \neq \mathbf{0}$. In terms of Fourier basis

$$q(\mathbf{x}) = \sum_{\mathbf{n} \in \mathbb{Z}^3} q_{\mathbf{n}} e^{2\pi i \mathbf{n}\cdot\mathbf{x}}, \qquad \text{where } q_{\mathbf{n}} = \int_\Omega q(\mathbf{x}) e^{-2\pi i \mathbf{n}\cdot\mathbf{x}}\, d\mathbf{x},$$

we see that

$$c_{\mathbf{k}}(q, q) = \sum_{\mathbf{n} \in \mathbb{Z}^3} |2\pi\mathbf{n} + \mathbf{k}|^2 q_{\mathbf{n}}^2 \| e^{2\pi i \mathbf{n}\cdot\mathbf{x}} \|^2.$$

Since $\mathbf{k} \in K = [-\pi, \pi]^3$ the term $|2\pi\mathbf{n} + \mathbf{k}|$ is zero only when $\mathbf{k} = \mathbf{n} = \mathbf{0}$, so for $\mathbf{k} \neq \mathbf{0}$ $c_{\mathbf{k}}(\cdot,\cdot)$ is coercive.

1.2.1 Mixed formulation

Problem 1.1. Eigenvalue problem in the mixed form ($\mathbf{k} \neq \mathbf{0}$).
Find triple $(\mathbf{u}, p, \lambda) \in \mathbf{H}_{\text{per}}(\text{curl}, \Omega) \times H^1_{\text{per}}(\Omega) \times \mathbb{R}$ s.t. $(\mathbf{u}, p) \neq (\mathbf{0}, 0)$ and for all $\mathbf{v} \in \mathbf{H}_{\text{per}}(\text{curl}, \Omega)$, $q \in H^1_{\text{per}}(\Omega)$

$$a_{\mathbf{k}}(\mathbf{u}, \mathbf{v}) + b_{\mathbf{k}}(\mathbf{v}, p) = \lambda\, m(\mathbf{u}, \mathbf{v}), \tag{1.19}$$

$$b_{\mathbf{k}}(\mathbf{u}, q) = 0. \tag{1.20}$$

This formulation is often used for theoretical investigations. It directly enforces the divergence condition and allows to work in a larger space $\mathbf{H}_{\mathrm{per}}(\mathrm{curl},\Omega)$, not in $\mathbf{V_k}$.

One can figure out that the solutions of the system actually have the form $(\mathbf{u}, 0, \lambda)$, where $\mathbf{u} \in \mathbf{V_k}$. The last one comes out from equation (1.20). To show that $p = 0$ let us put $\mathbf{v} = \nabla_{\mathbf{k}} p$ in equation (1.19), then we get $b_{\mathbf{k}}(\nabla_{\mathbf{k}} p, p) = c_{\mathbf{k}}(p,p) = 0$, what gives $p = 0$.

Let us define the solution operator $T\colon \mathbf{L}^2(\Omega) \to \mathbf{L}^2(\Omega)$ of Problem 1.1 as follows. For all $\mathbf{f} \in \mathbf{L}^2(\Omega)$, $T\mathbf{f} = \mathbf{u}$, where $\mathbf{u}$ is a solution of the following problem: find $(\mathbf{u}, p) \in \mathbf{H}_{\mathrm{per}}(\mathrm{curl},\Omega) \times H^1_{\mathrm{per}}(\Omega)$ s.t. for all $\mathbf{v} \in \mathbf{H}_{\mathrm{per}}(\mathrm{curl},\Omega)$, $q \in H^1_{\mathrm{per}}(\Omega)$

$$\begin{aligned} a_{\mathbf{k}}(\mathbf{u},\mathbf{v}) + b_{\mathbf{k}}(\mathbf{v},p) &= m(\mathbf{f},\mathbf{v}), \\ b_{\mathbf{k}}(\mathbf{u},q) &= 0. \end{aligned}$$

Lemma 1.4. *(see [8, Lemma 2])*
The operator T is compact and self-adjoint from $\mathbf{L}^2(\Omega)$ into itself.

As a conclusion, Problem 1.1 has an increasing sequence of eigenvalues

$$0 \le \lambda_1 \le \ldots \le \lambda_n \le \ldots$$

with finite dimensional eigenspaces.

1.2.2 Discretization

We apply the *Galerkin projection* method to construct a discrete approximation to Problem 1.1. Let $\mathbf{X}_h \subset \mathbf{H}_{\mathrm{per}}(\mathrm{curl},\Omega)$ and $Q_h \subset H^1_{\mathrm{per}}(\Omega)$ be finite dimensional subspaces. Define

$$\mathbf{V}_{h,\mathbf{k}} = \{\mathbf{v}_h \in \mathbf{X}_h \colon b_{\mathbf{k}}(\mathbf{v}_h, q_h) = 0 \quad \text{for all } q_h \in Q_h\}.$$

Note that $V_{h,\mathbf{k}}$ is not a subset of $V_{\mathbf{k}}$.

We require that the discrete spaces satisfy periodic boundary condition as the continuous spaces do. In practice it is done by identification of degrees of freedom, this will be explained later in Chapter 5.

Problem 1.2. Discrete form of Problem 1.1 ($\mathbf{k} \neq \mathbf{0}$).
Find a triple $(\mathbf{u}_h, p_h, \lambda_h) \in \mathbf{X}_h \times Q_h \times \mathbb{R}$ s.t. $(\mathbf{u}_h, p_h) \neq (\mathbf{0}, 0)$ and for all $\mathbf{v}_h \in \mathbf{X}_h$, $q_h \in Q_h$

$$\begin{aligned} a_{\mathbf{k}}(\mathbf{u}_h,\mathbf{v}_h) + b_{\mathbf{k}}(\mathbf{v}_h,p_h) &= \lambda_h\, m(\mathbf{u}_h,\mathbf{v}_h), \\ b_{\mathbf{k}}(\mathbf{u}_h,q_h) &= 0. \end{aligned}$$

The advantage of this formulation over one for the space $\mathbf{V}_{h,\mathbf{k}}$ is that we look for a solution in space $\mathbf{X}_h$, for which there exists *conforming finite elements* (will be explained later).

We say that a discrete form of the eigenvalue problem is a *spectrally correct approximation* of the original eigenvalue problem, if all eigenvectors and eigenvalues $(\mathbf{u}_h, \lambda_h)$ of the discrete form converge to the eigenvectors and eigenvalues $(\mathbf{u}, \lambda)$ of the original eigenvalue problem while $h \to 0$, and vice versa, all $(\mathbf{u}, \lambda)$ are approximated by $(\mathbf{u}_h, \lambda_h)$ respecting their multiplicity. We want Problem 1.2 to be a spectrally correct approximation of Problem 1.1. In order to analyze the convergence of the discrete eigenvalue solutions to the continuous ones we apply the abstract theory developed in [7] and [6].

By analogy with the solution operator T of Problem 1.1 we define the discrete solution operator $T_h\colon \mathbf{L}^2(\Omega) \to \mathbf{X}_h$ of Problem 1.2 as follows. For all $\mathbf{f} \in \mathbf{L}^2(\Omega)$, $T_h\mathbf{f} = \mathbf{u}_h \in \mathbf{X}_h$, where $\mathbf{u}_h$ is from the following problem.

Find $(\mathbf{u}_h, p_h) \in \mathbf{X}_h \times Q_h$ s.t. for all $\mathbf{v}_h \in \mathbf{X}_h$, $q_h \in Q_h$

$$\begin{aligned} a_{\mathbf{k}}(\mathbf{u}_h,\mathbf{v}_h) + b_{\mathbf{k}}(\mathbf{v}_h,p_h) &= m(\mathbf{f},\mathbf{v}_h), \\ b_{\mathbf{k}}(\mathbf{u}_h,q_h) &= 0. \end{aligned}$$

Theorem 1.5. *(see [8, Theorem 2])*
If the spaces $\mathbf{X}_h$, Q_h, $\mathbf{V}_{h,\mathbf{k}}$ satisfy the conditions below, then the sequence T_h converges uniformly to T in $\mathcal{L}(\mathbf{L}^2(\Omega), \mathbf{H}_{\text{per}}(\text{curl}, \Omega))$, i.e. there exists $\rho_3(h)$, tending to zero as $h \to 0$ s.t.

$$\|T\mathbf{f} - T_h\mathbf{f}\|_{\text{curl}} \leq \rho_3(h)\|\mathbf{f}\|_0 \qquad \text{for all } \mathbf{f} \in \mathbf{L}^2(\Omega).$$

The conditions are:

1. *Ellipticity on $\mathbf{V}_{h,\mathbf{k}}$*
   There exists $C > 0$ s.t.

$$a_{\mathbf{k}}(\mathbf{u}_h, \mathbf{u}_h) \geq C\|\mathbf{u}_h\|^2_{\mathbf{L}^2} \qquad \text{for all } \mathbf{u}_h \in \mathbf{V}_{h,\mathbf{k}},$$

2. *Weak approximability of $H^1_{\text{per}}(\Omega)$*
   There exists $\rho_1(h) > 0$, tending to zero as $h \to 0$ s.t.

$$\sup_{\mathbf{v}_h \in \mathbf{V}_{h,\mathbf{k}}} \frac{b_{\mathbf{k}}(\mathbf{v}_h, q)}{\|\mathbf{v}_h\|_{\text{curl}}} \leq \rho_1(h)\|q\|_{H^1} \qquad \text{for all } q \in H^1_{\text{per}}(\Omega),$$

3. *Strong approximability of $\mathbf{V}_{\mathbf{k}}$*
   For some $r > 0$ there exists $\rho_2(h) > 0$, tending to zero as $h \to 0$ s.t. for any $\mathbf{u} \in \mathbf{V}_{\mathbf{k}} \cap (\mathbf{H}^{1+r}(\Omega))^3$ there exists $\mathbf{u}_h \in \mathbf{V}_{h,\mathbf{k}}$ satisfying

$$\|\mathbf{u} - \mathbf{u}_h\|_{\text{curl}} \leq \rho_2(h)\|\mathbf{u}\|_{\mathbf{H}^{1+r}}.$$

Here we should recall the result of Lemma 1.4, that T is a compact and self-adjoint operator. For such an operator, the uniform convergence $T_h \to T$ in the operator norm, which is a result of Theorem 1.5, is a sufficient and necessary condition to have discrete convergence of the spectrum. It is formulated in the next theorem.

Theorem 1.6. *(see [8, Theorem 3])*
There exists a constant C s.t. for all $\mathbf{f} \in \mathbf{L}^2(\Omega)$ it holds

$$\|T\mathbf{f} - T_h\mathbf{f}\|_{\text{curl}} \leq Ch^t\|\mathbf{f}\|_0,$$

where $t > 0$ depends on the regularity of the material distribution ε.

Let λ_j be an eigenvalue of Problem 1.1 with multiplicity m_j, and denote by E_j the corresponding eigenspace, then exactly m_j discrete eigenvalues $\lambda_{h,j_1}, \dots, \lambda_{h,j_{m_j}}$ of Problem 1.2 converge to λ_j. Moreover, for some $h < h_0$

$$\left|\frac{1}{m_j}\sum_{n=1}^{m_j} \lambda_{h,j_n} - \lambda_j\right| \leq Ch^{2t},$$
$$\text{dist}(\text{span}\{\mathbf{u}_{h,j_1}, \dots, \mathbf{u}_{h,j_{m_j}}\}, E_j) \leq Ch^t.$$

As we see, Theorems 1.5 and 1.6 state that under the proper conditions Problem 1.2 is a spectrally correct approximation of Problem 1.1.

1.3 Computer-assisted proof for band gap

Theorem 1.6 states that for a given $\mathbf{k}$ one can obtain approximations to the exact eigenvalues. So, defining a mesh in K one can compute an approximated band structure for a given material distribution and then look for band gaps. It may happen that some band gaps we observe in the approximated band structure for a finite set of $\mathbf{k}$ will disappear after adding some more $\mathbf{k}$-points or improving precision of the eigenvalue computations. So what we observe in the approximated band structure is just a hint at real band gap and we need a theory to prove that it does exist.

In the paper [19] it is presented a way to prove existence of a band gap using eigenvalue bounds for a fixed $\mathbf{k}$ and a perturbation analysis to extend these bounds between nodes of a finite mesh in K. In such a way it is possible to prove existence of true band gaps.

Note that the result in [19] was obtained for the 2D case of polarized waves, where the eigenvalue problem is solved in $H^1_{\text{per}}(\Omega)$. Therefore, the result cannot be directly applied to our more general case of $\mathbf{H}_{\text{per}}(\text{curl}, \Omega)$. More precisely, the eigenvalue bounds for a fixed $\mathbf{k}$ are not valid, but similar perturbation argument still can be used. The 3D case will be considered in [5].

First, we need to find a material distribution which exhibits a "numerical" band gap for a finite set of $\mathbf{k}$-points. Technically it means that we look for a periodic function ε for which Problem 1.2 gives eigenvalue approximations $\lambda_{\mathbf{k},h,j}$, $j \in \{1, \dots, n\}$ and $\mathbf{k} \in \mathcal{K}$, where $\mathcal{K}$ is a grid in K. Moreover, we assume there exists l s.t. $\max_{\mathbf{k}\in\mathcal{K}}\{\lambda_{\mathbf{k},h,l}\} < \min_{\mathbf{k}\in\mathcal{K}}\{\lambda_{\mathbf{k},h,l+1}\}$.

Second, one applies a numerical procedure to obtain bounds for the exact eigenvalues $\lambda_{\mathbf{k},j}$, let us denote them $\underline{\lambda}_{\mathbf{k},j} \le \lambda_{\mathbf{k},j} \le \overline{\lambda}_{\mathbf{k},j}$, for any $j \in \{1, \dots, n\}$ and $\mathbf{k} \in \mathcal{K}$. As it was already mentioned, [19] describes an algorithm how to get such bounds for the 2D case of polarized waves. For our 3D case there is no algorithm yet, nevertheless let us assume that $\underline{\lambda}_{\mathbf{k},j}$ and $\overline{\lambda}_{\mathbf{k},j}$ are available. In practice one can obtain reasonably precise estimates by considering h-convergence of $\lambda_{\mathbf{k},h,j}$ on a sequence of embedded meshes. If the eigenvalues converge monotonically (usually monotonically decreasing), then one bound is the value on the finest mesh, another bound can be well estimated by extrapolating the h-convergence. After that stage assume that we confirmed the gap and $\max_{\mathbf{k}\in\mathcal{K}}\{\overline{\lambda}_{\mathbf{k},l}\} < \min_{\mathbf{k}\in\mathcal{K}}\{\underline{\lambda}_{\mathbf{k},l+1}\}$.

Third, we use the a perturbation argument. We know that $\lambda \in (\overline{\lambda}_{\mathbf{k},l}, \underline{\lambda}_{\mathbf{k},l+1})$ is not an eigenvalue of Problem 1.1 for any $\mathbf{k} \in \mathcal{K}$. Let us consider perturbation of the eigenvalues when $\mathbf{k}$ has a small variation $\mathbf{h}$. In case $\mathcal{K}$ is "sufficiently dense" in K, we may expect that λ is not in the spectrum for any $\mathbf{k} \in K$.

For simplicity let us denote $X = \mathbf{H}_{\text{per}}(\text{curl}, \Omega)$, $H = \mathbf{L}^2(\Omega)$. They build a Gelfand triple $X \subset H \subset X'$.

For $\delta > 0$ the shifted operator $A^\delta_{\mathbf{k}} \colon X \to X'$ is defined by

$$\langle A^\delta_{\mathbf{k}}\mathbf{u}, \mathbf{v}\rangle = (\varepsilon^{-1} \nabla_{\mathbf{k}} \times \mathbf{u}, \nabla_{\mathbf{k}} \times \mathbf{v})_{\mathbf{L}^2} + \delta(\mathbf{u}, \mathbf{v})_{\mathbf{L}^2}.$$

The spectrum of the operator $A^\delta_{\mathbf{k}}$ is the spectrum of the operator $A_{\mathbf{k}}$ shifted with δ to the right.

We fix $\mathbf{k} \in \mathcal{K}$ and δ. Let us define a scaled norm $\|\cdot\|_H$ and the associated energy norm $\|\cdot\|_X$

$$\|\mathbf{u}\|_H = \sqrt{\delta}\|\mathbf{u}\|_{\mathbf{L}^2}, \qquad \|\mathbf{u}\|_X = \sqrt{\|\varepsilon^{-\frac{1}{2}} \nabla_{\mathbf{k}} \times \mathbf{u}\|^2_{\mathbf{L}^2} + \|\mathbf{u}\|^2_H}.$$

In terms of the energy norm we have

$$\langle A^\delta_{\mathbf{k}}\mathbf{u}, \mathbf{u}\rangle \ge \|\mathbf{u}\|^2_X$$
$$\langle A^\delta_{\mathbf{k}}\mathbf{u}, \mathbf{v}\rangle \le \|\mathbf{u}\|_X\|\mathbf{v}\|_X.$$

This shows that $A^\delta_{\mathbf{k}} \colon X \to X'$, which is isometric (in the energy norm) and, in particular, $(A^\delta_{\mathbf{k}})^{-1} \colon X' \to X$ exists and $\|(A^\delta_{\mathbf{k}})^{-1}\|_{L(X',X)} \le 1$.

Denote the embedding $E \colon X \to H$, then $E' \colon H \to X'$. Since $\|\mathbf{u}\|_H \le \|\mathbf{u}\|_X$ we have $\|E\|_{L(X,H)} \le 1$ and so $\|E'\|_{L(H,X')} \le 1$. Introduce an auxiliary operator

$$B_{\mathbf{k}} = E(A^\delta_{\mathbf{k}})^{-1}E' \colon H \to H.$$

We have the spectrum $\sigma(B_{\mathbf{k}}) = \{\frac{1}{\lambda+\delta} \mid \lambda \in \sigma(A_{\mathbf{k}})\}$. The interval between the eigenvalues $[\lambda_{\mathbf{k},l}, \lambda_{\mathbf{k},l+1}]$ for the operator $A_{\mathbf{k}}$ translates to the interval $[\mu_{\mathbf{k},l+1}, \mu_{\mathbf{k},l}]$ for the operator $B_{\mathbf{k}}$, where $\mu_{\mathbf{k},l+1} = \frac{1}{\lambda_{\mathbf{k},l+1}+\delta}$ and $\mu_{\mathbf{k},l} = \frac{1}{\lambda_{\mathbf{k},l}+\delta}$.

Theorem 1.7. *Perturbation theorem for band gap.*
Let $B(\mathbf{k}, r) = \{\mathbf{k}' \in \mathbb{R}^3 \mid |\mathbf{k}' - \mathbf{k}| < r\}$ and $\varepsilon_{\min} = \min_{\mathbf{x}\in\Omega} \varepsilon(\mathbf{x})$. Suppose that for the operator $A_{\mathbf{k}}$, there exists an interval $[a, b]$ s.t. for some $l \in \mathbb{N}$

1. *$[a, b] \subset (\overline{\lambda}_{\mathbf{k},l}, \underline{\lambda}_{\mathbf{k},l+1})$ for all $\mathbf{k} \in \mathcal{K}$,*
2. *$K \subset \bigcup_{\mathbf{k}\in\mathcal{K}} B(\mathbf{k}, r_{\mathbf{k}})$, where $r_{\mathbf{k}}$ holds*

$$r_{\mathbf{k}} < \frac{\beta_{\mathbf{k}}\sqrt{\delta\varepsilon_{\min}}}{\sqrt{(1+\beta_{\mathbf{k}})(1+2\beta_{\mathbf{k}})}},$$

$$\beta_{\mathbf{k}} = \min\left\{\frac{a-\lambda_{\mathbf{k},l}}{(a+\delta)(\lambda_{\mathbf{k},l}+\delta)}, \frac{\lambda_{\mathbf{k},l+1}-b}{(b+\delta)(\lambda_{\mathbf{k},l+1}+\delta)}\right\}.$$

Then $[a, b]$ is contained in the spectral gap, i.e. $[a, b] \subset (\overline{\lambda}_{\mathbf{k},l}, \underline{\lambda}_{\mathbf{k},l+1})$ for all $\mathbf{k} \in K$.

Proof. Define the resolvent

$$R_{\mathbf{k}}(\mu) = (B_{\mathbf{k}} - \mu I)^{-1} \colon H \to H.$$

One may show that if $\mu \in [\mu_{\mathbf{k},l+1} + \beta_{\mathbf{k}}, \mu_{\mathbf{k},l} - \beta_{\mathbf{k}}]$, then for any $\mathbf{k}$ there exists $R_{\mathbf{k}}(\mu)$ and $\|R_{\mathbf{k}}(\mu)\|_H \le \beta_{\mathbf{k}}^{-1}$. Expanding $\mathbf{u} \in H$ with respect to a complete orthonormal system of eigenfunctions $\{\mathbf{u}_{\mathbf{k},n}\}_{n\in\mathbb{N}}$ of $B_{\mathbf{k}}$ we estimate

$$\|R_{\mathbf{k}}(\mu)\mathbf{u}\|_H^2 = \sum_{n\in\mathbb{N}} \frac{1}{(\mu_{\mathbf{k},n} - \mu)^2} |\langle \mathbf{u}, \mathbf{u}_{\mathbf{k},n}\rangle|^2 \le \frac{1}{\beta_{\mathbf{k}}^2}\|\mathbf{u}\|_H^2.$$

Now we consider $\mathbf{h}$ s.t. $|\mathbf{h}| < r_{\mathbf{k}}$, some small perturbation of $\mathbf{k}$. We want to show that $R_{\mathbf{k}+\mathbf{h}}(\mu)$ exists. First, let us prove the following representation

$$R_{\mathbf{k}+\mathbf{h}}(\mu) = R_{\mathbf{k}}(\mu)\Big(I + (B_{\mathbf{k}+\mathbf{h}} - B_{\mathbf{k}})R_{\mathbf{k}}(\mu)\Big)^{-1}. \tag{1.21}$$

Simple calculations show that

$$\begin{aligned}(B_{\mathbf{k}+\mathbf{h}} - \mu I) &= (B_{\mathbf{k}+\mathbf{h}} - \mu I) + (B_{\mathbf{k}} - \mu I) - (B_{\mathbf{k}} - \mu I)\\ &= (B_{\mathbf{k}} - \mu I)\Big(I + (B_{\mathbf{k}+\mathbf{h}} - B_{\mathbf{k}})(B_{\mathbf{k}} - \mu I)^{-1}\Big).\end{aligned}$$

Inverse of this gives us (1.21).

If $R_{\mathbf{k}}(\mu)$ exists, then it is enough to show that

$$\|(B_{\mathbf{k}+\mathbf{h}} - B_{\mathbf{k}})R_{\mathbf{k}}(\mu)\|_H < 1, \tag{1.22}$$

hence the right hand term of (1.21) is invertible, and so $R_{\mathbf{k}+\mathbf{h}}(\mu)$ exists.

Define $S_{\mathbf{k},\mathbf{h}} \colon X \to X'$

$$\langle S_{\mathbf{k},\mathbf{h}}\mathbf{u}, \mathbf{v}\rangle = s_{\mathbf{k},\mathbf{h}}(\mathbf{u}, \mathbf{v}) = \langle A^{\delta}_{\mathbf{k}+\mathbf{h}}\mathbf{u}, \mathbf{v}\rangle - \langle A^{\delta}_{\mathbf{k}}\mathbf{u}, \mathbf{v}\rangle.$$

We want to estimate $\|S_{\mathbf{k},\mathbf{h}}\|_{L(X,X')} \leq C|\mathbf{h}|$.

$$
\begin{aligned}
|s_{\mathbf{k},\mathbf{h}}(\mathbf{u},\mathbf{v})| =& |\mathrm{i}(\varepsilon^{-1}\mathbf{h}\times\mathbf{u}, \nabla_{\mathbf{k}}\times\mathbf{v})_{\mathbf{L}^2} - \mathrm{i}(\varepsilon^{-1}\nabla_{\mathbf{k}}\times\mathbf{u}, \mathbf{h}\times\mathbf{v})_{\mathbf{L}^2} + \\
&+ (\varepsilon^{-1}\mathbf{h}\times\mathbf{u}, \mathbf{h}\times\mathbf{v})_{\mathbf{L}^2}| \\
\leq& \|\varepsilon^{-\frac{1}{2}}\mathbf{h}\times\mathbf{u}\|_{\mathbf{L}^2}\|\varepsilon^{-\frac{1}{2}}\nabla_{\mathbf{k}}\times\mathbf{v}\|_{\mathbf{L}^2} + \|\varepsilon^{-\frac{1}{2}}\nabla_{\mathbf{k}}\times\mathbf{u}\|_{\mathbf{L}^2}\|\varepsilon^{-\frac{1}{2}}\mathbf{h}\times\mathbf{v}\|_{\mathbf{L}^2} + \\
&+ \|\varepsilon^{-1}\mathbf{h}\times\mathbf{u}\|_{\mathbf{L}^2}\|\mathbf{h}\times\mathbf{v}\|_{\mathbf{L}^2} \\
\leq& \varepsilon_{\min}^{-\frac{1}{2}}|\mathbf{h}|\Big(\|\mathbf{u}\|_{\mathbf{L}^2}\|\varepsilon^{-\frac{1}{2}}\nabla_{\mathbf{k}}\times\mathbf{v}\|_{\mathbf{L}^2} + \|\varepsilon^{-\frac{1}{2}}\nabla_{\mathbf{k}}\times\mathbf{u}\|_{\mathbf{L}^2}\|\mathbf{v}\|_{\mathbf{L}^2} + \\
&+ \varepsilon_{\min}^{-\frac{1}{2}}|\mathbf{h}|\|\mathbf{u}\|_{\mathbf{L}^2}\|\mathbf{v}\|_{\mathbf{L}^2}\Big).
\end{aligned}
$$

The last sum in the parenthesis can be represented and estimates as

$$
\begin{aligned}
ad + be + cf &= (a,b,c)(d,e,f)^\top = (\alpha a, b, \sqrt{\alpha}c)(\frac{1}{\alpha}d, e, \frac{1}{\sqrt{\alpha}}f)^\top \\
&= \frac{1}{\alpha}(\alpha a, b, \sqrt{\alpha}c)(d, \alpha e, \sqrt{\alpha}f)^\top \leq \frac{1}{\alpha}|(\alpha a, b, \sqrt{\alpha}c)|\cdot|(d, \alpha e, \sqrt{\alpha}f)|,
\end{aligned}
$$

where $\alpha > 0$ is some number. If we use this representation to estimate $|s_{\mathbf{k},\mathbf{h}}(\mathbf{u},\mathbf{v})|$, it follows that

$$
\begin{aligned}
|s_{\mathbf{k},\mathbf{h}}(\mathbf{u},\mathbf{v})| \leq& \frac{1}{\alpha}\varepsilon_{\min}^{-\frac{1}{2}}|\mathbf{h}|\sqrt{\|\varepsilon^{-\frac{1}{2}}\nabla_{\mathbf{k}}\times\mathbf{u}\|_{\mathbf{L}^2}^2 + \alpha(\alpha + \varepsilon_{\min}^{-\frac{1}{2}}|\mathbf{h}|)\|\mathbf{u}\|_{\mathbf{L}^2}^2}\cdot \\
&\sqrt{\|\varepsilon^{-\frac{1}{2}}\nabla_{\mathbf{k}}\times\mathbf{v}\|_{\mathbf{L}^2}^2 + \alpha(\alpha + \varepsilon_{\min}^{-\frac{1}{2}}|\mathbf{h}|)\|\mathbf{v}\|_{\mathbf{L}^2}^2},
\end{aligned}
$$

what gives us

$$
|s_{\mathbf{k},\mathbf{h}}(\mathbf{u},\mathbf{v})| \leq \frac{1}{\alpha}\varepsilon_{\min}^{-\frac{1}{2}}|\mathbf{h}|\|\mathbf{u}\|_X\|\mathbf{v}\|_X, \tag{1.23}
$$

if

$$
\alpha(\alpha + \varepsilon_{\min}^{-\frac{1}{2}}|\mathbf{h}|) \leq \delta. \tag{1.24}
$$

Solving (1.24) one obtains

$$
\alpha \leq \frac{\sqrt{q^2+4\delta}-q}{2}, \quad \text{where } q = \varepsilon_{\min}^{-\frac{1}{2}}|\mathbf{h}|.
$$

Since we want a large α, we take the boundary and put it into (1.23) and so obtain the final estimation

$$
|s_{\mathbf{k},\mathbf{h}}(\mathbf{u},\mathbf{v})| \leq \frac{2q}{\sqrt{q^2+4\delta}-q}\|\mathbf{u}\|_X\|\mathbf{v}\|_X, \quad \text{where } q = \varepsilon_{\min}^{-\frac{1}{2}}|\mathbf{h}|.
$$

Let us estimate $\|B_{\mathbf{k}+\mathbf{h}} - B_{\mathbf{k}}\|_H$

$$
\begin{aligned}
B_{\mathbf{k}+\mathbf{h}} - B_{\mathbf{k}} &= E\left((A_{\mathbf{k}}^\delta + S_{\mathbf{k},\mathbf{h}})^{-1} - (A_{\mathbf{k}}^\delta)^{-1}\right)E' \\
&= E\left((I - (A_{\mathbf{k}}^\delta)^{-1}S_{\mathbf{k},\mathbf{h}})^{-1} - I\right)(A_{\mathbf{k}}^\delta)^{-1}E' \\
&= E\left(\sum_{n\geq 1}((A_{\mathbf{k}}^\delta)^{-1}S_{\mathbf{k},\mathbf{h}})^n\right)(A_{\mathbf{k}}^\delta)^{-1}E'.
\end{aligned}
$$

The sum can be estimated by Neumann series, which is converging due to the assumptions of the theorem

$$
c = \|(A_{\mathbf{k}}^\delta)^{-1}S_{\mathbf{k},\mathbf{h}}\|_{L(X,X)} \leq \|(A_{\mathbf{k}}^\delta)^{-1}\|_{L(X',X)}\|S_{\mathbf{k},\mathbf{h}}\|_{L(X,X')} \leq \frac{2q}{\sqrt{q^2+4\delta}-q} < 1,
$$

and finally it gives

$$\|B_{\mathbf{k}+\mathbf{h}} - B_{\mathbf{k}}\|_H \le \|E\|_{L(X,H)} \left(\frac{c}{1-c}\right) \|(A_{\mathbf{k}}^{\delta})^{-1}\|_{L(X',X)} \|E'\|_{L(H,X')}.$$

Recalling the estimates for E, E', $(A_{\mathbf{k}}^{\delta})^{-1}$, $S_{\mathbf{k},\mathbf{h}}$ and $R_{\mathbf{k}}(\mu)$ we get

$$\|B_{\mathbf{k}+\mathbf{h}} - B_{\mathbf{k}}\|_H \|R_{\mathbf{k}}(\mu)\|_H \le \left(\frac{1}{1-c} - 1\right) \frac{1}{\beta_{\mathbf{k}}}. \tag{1.25}$$

Now we show that the assumptions of the theorem provide (1.25) can be estimated by 1, then according to (1.21) and (1.22) the resolvent $R_{\mathbf{k}+\mathbf{h}}(\mu)$ exists for any $\mu \in [\mu_{\mathbf{k},l+1} + \beta_{\mathbf{k}}, \mu_{\mathbf{k},l} - \beta_{\mathbf{k}}]$, what follows that the interval $[\mu_{\mathbf{k},l+1} + \beta_{\mathbf{k}}, \mu_{\mathbf{k},l} - \beta_{\mathbf{k}}]$ is excluded from the spectrum of $B_{\mathbf{k}+\mathbf{h}}$.

$\mu \in [\mu_{\mathbf{k},l+1} + \beta_{\mathbf{k}}, \mu_{\mathbf{k},l} - \beta_{\mathbf{k}}]$ corresponds to $\lambda \in [a, b]$ for the operator $A_{\mathbf{k}+\mathbf{h}}$, where

$$a = \frac{1}{\mu_{\mathbf{k},l} - \beta_{\mathbf{k}}} - \delta, \qquad b = \frac{1}{\mu_{\mathbf{k},l+1} + \beta_{\mathbf{k}}} - \delta.$$

Then $\beta_{\mathbf{k}}$ can be represented as

$$\beta_{\mathbf{k}}(a) = \mu_{\mathbf{k},l} - \frac{1}{a+\delta}, \qquad \beta_{\mathbf{k}}(b) = \frac{1}{b+\delta} - \mu_{\mathbf{k},l+1}.$$

At last, in terms of λ it is

$$\beta_{\mathbf{k}}(a) = \frac{a - \lambda_{\mathbf{k},l}}{(a+\delta)(\lambda_{\mathbf{k},l} + \delta)}, \qquad \beta_{\mathbf{k}}(b) = \frac{\lambda_{\mathbf{k},l+1} - b}{(b+\delta)(\lambda_{\mathbf{k},l+1} + \delta)}.$$

From the inequality

$$\left(\frac{1}{1-c} - 1\right) \frac{1}{\beta_{\mathbf{k}}} < 1$$

one derives

$$\frac{2q}{\sqrt{q^2 + 4\delta} - q} < \frac{\beta_{\mathbf{k}}}{1 + \beta_{\mathbf{k}}}, \qquad \beta_{\mathbf{k}} = \min\{\beta_{\mathbf{k}}(a), \beta_{\mathbf{k}}(b)\},$$

then

$$q < \frac{\sqrt{\delta}\beta_{\mathbf{k}}}{\sqrt{(1+\beta_{\mathbf{k}})(1+2\beta_{\mathbf{k}})}},$$

and the final formula for allowed perturbation is

$$|\mathbf{h}| < \frac{\beta_{\mathbf{k}}\sqrt{\delta\varepsilon_{\min}}}{\sqrt{(1+\beta_{\mathbf{k}})(1+2\beta_{\mathbf{k}})}}.$$

Since the previous considerations are true for any $\mathbf{k} \in \mathcal{K}$ and the system of balls $B(\mathbf{k}, r_{\mathbf{k}})$ cover the whole K we conclude that the interval $[a, b]$ is excluded from the spectrum of $A_{\mathbf{k}}$ for any $\mathbf{k} \in K$.

□

Chapter 2

Finite Elements

2.1 Standard lowest order elements

We are going to recall only necessary facts concerning finite elements. For further details we refer to standard text books e.g. [28], but many important facts are also collected in [17].

Definition 2. The *Finite Element Method* (FEM) is a Galerkin method which is characterized by the following principles in the construction of a discrete subspace $\mathbf{X}_h$:

1. The domain $\bar{\Omega}$ is represented as a finite union of non-overlapping polyhedral elements Ω_c.
2. $\mathbf{X}_h$ consists of piecewise polynomials, so that the restriction of $\mathbf{X}_h$ onto an element Ω_c is a polynomial space.
3. $\mathbf{X}_h$ has a basis consisting of functions with local supports, i.e. the functions are non-zero only on few elements.

Definition 3. A *finite element* is the triplet $(\Omega_c, \mathcal{P}_c, \Sigma_c)$, where

- $\Omega_c \subset \mathbb{R}^d$ is the element domain, a bounded closed set with non-empty interior and piecewise smooth boundary,
- $\mathcal{P}_c$ is the space of *shape functions*, a finite-dimensional space of functions on Ω_c,
- Σ_c is the set of *degrees of freedom*, a basis of $\mathcal{P}_c'$ (the dual space).

Let $\bar{\Omega} = \bigcup_{c \in \mathcal{C}_h} \bar{\Omega}_c$ be a decomposition of the domain into axis-parallel hexahedral cells (bricks) $c \in \mathcal{C}_h$. In general all our results can be extended to unstructured meshes of tetrahedrons, we fix the hexahedral elements just as an instance. Let $\hat{\Omega} = [0,1]^3$ be the *reference element*, $\varphi_c \colon \hat{\Omega} \to \Omega_c$ is the *mapping* from the reference element to a *physical element*, φ_c is a C^1 one-to-one and onto map. The *Jacobian matrix* of φ_c is defined by

$$F_c(\hat{\mathbf{x}}) = \left(\frac{\partial \varphi_{c,i}(\hat{\mathbf{x}})}{\partial \hat{\mathbf{x}}_j} \right)_{i,j=1,\dots,d},$$

and the *Jacobian determinant* is $J_c(\hat{\mathbf{x}}) = \det(F_c(\hat{\mathbf{x}}))$.

Let us define the simplest $\mathbf{H}(\mathrm{curl})$- and H^1-conforming finite elements

$$\begin{aligned} \mathbf{X}_{h,0} &= \{\mathbf{u} \in \mathbf{H}_{\mathrm{per}}(\mathrm{curl}, \Omega) \colon F_c^T \mathbf{u} \circ \varphi_c \in \mathcal{Q}^{0,1,1} \times \mathcal{Q}^{1,0,1} \times \mathcal{Q}^{1,1,0} \text{ for all } c \in \mathcal{C}_h\}, \\ Q_{h,0} &= \{q \in H^1_{\mathrm{per}}(\Omega) \colon q \circ \varphi_c \in \mathcal{Q}^{1,1,1} \text{ for all } c \in \mathcal{C}_h\}, \end{aligned}$$

where $Q^{l,n,m}$ is a polynomial space

$$Q^{l,n,m} = \left\{ p(x,y,z) = \sum_{p=0}^{l}\sum_{j=0}^{n}\sum_{k=0}^{m} c_{pjk} x^p y^j z^k \right\}.$$

Let $\mathcal{V}_h$ be the set of all vertices v with the coordinates $\mathbf{z}_v \in \bar{\Omega}$, $\mathcal{E}_h$ be the set of all edges e identified with the two ordered vertices $(\mathbf{x}_e, \mathbf{y}_e) \subset \bar{\Omega}$. The edge has the midpoint $\mathbf{m}_e = 0.5(\mathbf{x}_e + \mathbf{y}_e)$, the unit tangent vector $\mathbf{t}_e = (\mathbf{y}_e - \mathbf{x}_e)/\|\mathbf{y}_e - \mathbf{x}_e\|$ and the edge curve itself is denoted by $\Gamma_e = \mathrm{conv}\{\mathbf{x}_e, \mathbf{y}_e\}$.

Finite elements $\mathbf{X}_{h,0}$ have edge-based degrees of freedom, for $\mathbf{u} \in \mathbf{H}_{\mathrm{per}}(\mathrm{curl}, \Omega)$ s.t. there also exists $\mathbf{u}|_{\Gamma_e} \in \mathbf{L}^2(\Gamma_e)$ define

$$\ell_{e,0}(\mathbf{u}) = \int_{\mathbf{x}_e}^{\mathbf{y}_e} \mathbf{u} \cdot \mathbf{t}_e ds \qquad \text{for all } e \in \mathcal{E}_h.$$

Finite elements $Q_{h,0}$ have vertex-based degrees of freedom, for a sufficiently smooth $q \in H^1_{\mathrm{per}}(\Omega)$ define

$$\ell_{v,0}(q) = q(\mathbf{z}_v) \qquad \text{for all } v \in \mathcal{V}_h.$$

The finite elements $\mathbf{X}_{h,0}$ are called the lowest-order Nédélec elements of the first kind and $Q_{h,0}$ are called the linear elements.

There exists the nodal basis $\{\psi_{e,0} \colon e \in \mathcal{E}_h\}$ of $\mathbf{X}_{h,0}$ dual to $\{\ell_{e,0} \colon e \in \mathcal{E}_h\}$ and the nodal basis $\{\phi_{v,0} \colon v \in \mathcal{V}_h\}$ of $Q_{h,0}$ dual to $\{\ell_{v,0} \colon v \in \mathcal{V}_h\}$. The basis functions and degrees of freedom are identified with so-called *nodal points* which are the midpoints of edges $\{\mathbf{m}_e\}$ and the vertices $\{\mathbf{z}_v\}$, respectively.

For $\mathbf{u}$ and q, which satisfy the assumptions for the degrees of freedom, we define interpolation operators to the finite element spaces

$$\Pi_{\mathbf{X}_{h,0}}(\mathbf{u}) = \sum_{e \in \mathcal{E}_h} \ell_{e,0}(\mathbf{u})\psi_{e,0},$$
$$\Pi_{Q_{h,0}}(q) = \sum_{v \in \mathcal{V}_h} \ell_{v,0}(q)\phi_{v,0}.$$

Since the definition uses the nodal bases, one may check that $(\Pi_{\mathbf{X}_{h,0}})^2 = \Pi_{\mathbf{X}_{h,0}}$ and $(\Pi_{Q_{h,0}})^2 = \Pi_{Q_{h,0}}$. It means that the interpolation operators are projectors.

In practice the global finite element space ($\mathbf{X}_{h,0}$ or $Q_{h,0}$) is assembled in the following way. There is the reference finite element, a local polynom space formed of shape functions defined on the reference element $\hat{\Omega}$. With help of φ_c the shape functions are mapped to a physical element Ω_c and so form a finite element there. Altogether these elements form the global finite element space. The concrete shape functions we use will be given later in Section 2.2.

It is important to do the mapping properly to guarantee that the finite element spaces are subspaces of the related Sobolev spaces.

Lemma 2.1. *(see [28, Chapter 5])*
For $\hat{\mathbf{u}} \in \mathbf{H}(\mathrm{curl}, \hat{\Omega})$ and $\hat{q} \in H^1(\Omega)$ the conforming mapping is given by the following formulas

$$\begin{aligned}
\mathbf{u} &= F_c^{-T}\hat{\mathbf{u}} \circ \varphi_c^{-1} \in \mathbf{H}(\mathrm{curl}, \hat{\Omega}_c),\\
\nabla_{\mathbf{x}} \times \mathbf{u} &= J_c^{-1} F_c \nabla_{\hat{\mathbf{x}}} \times \hat{\mathbf{u}} \circ \varphi_c^{-1},\\
q &= \hat{q} \circ \varphi_c^{-1} \in H^1(\Omega_c),\\
\nabla_{\mathbf{x}} q &= F_c^{-T} \nabla_{\hat{\mathbf{x}}} \hat{q} \circ \varphi_c^{-1}.
\end{aligned}$$

For the implementation of the finite element method one needs a numerical integration, which is used for matrix assembling. In this thesis we always assume an exact numerical integration.

Although the standard elements work well for the original Maxwell's equations, they cannot be directly used for our formulation with the k-shifted operators. To illustrate why, we give a simple example.

For simplicity let us consider the reference finite element only

$$Q_h = \mathcal{Q}^{1,1,1}, \qquad \mathbf{X}_h = \mathcal{Q}^{0,1,1} \times \mathcal{Q}^{1,0,1} \times \mathcal{Q}^{1,1,0}.$$

As a discrete analog of Corollary 1.2 we would like to have that $\nabla_{\mathbf{k}} Q_h$ is contained in $\mathbf{X}_h$ since $\nabla_{\mathbf{k}} H^1_{\mathrm{per}}(\Omega) \subset \mathbf{H}_{\mathrm{per}}(\mathrm{curl}, \Omega)$. But for $\mathbf{k} \neq \mathbf{0}$ $(\nabla + i\mathbf{k})\mathcal{Q}^{1,1,1}$ is not contained in $\mathbf{X}_h$, because $\mathbf{k}\mathcal{Q}^{1,1,1} \not\subseteq \mathcal{Q}^{0,1,1} \times \mathcal{Q}^{1,0,1} \times \mathcal{Q}^{1,1,0}$.

That is why we need to use special elements with k-shifted basis functions, which allow to construct the discrete exact sequence with respect to the operator $\nabla_{\mathbf{k}}$.

2.2 High order finite elements

Although Problem 1.2 can be successfully solved with the help of the lowest order finite elements, it may not be efficient. If the solution is assumed to have some additional regularity, one can expect that high order finite elements provide higher convergence rate and are more efficient in practice.

In construction of high order finite elements we follow an approach presented in [41]. The approach has the following advantages.

- It provides a sequence of *hierarchical finite element spaces* with an arbitrary and variable polynomial order. This flexibility is especially needed for a *hp-adaptive* method.
- Shape functions for the $\mathbf{H}(\mathrm{curl})$-conforming finite element space explicitly includes gradient functions. This fact allows us to construct an inexpensive discrete gradient operator, which we need in Sections 3.2 and 4.3.
- The resulting finite element space consists of well-structured subspaces. The *exact sequence property* (Lemma 2.5) is enforced and even more, the *local exact sequence property* exists. The property can be used e.g. for the construction of an efficient multigrid preconditioner.
- The construction of shape functions is done recursively by using orthogonal polynomials and element-based spatial variables. It allows a simple implementation as well as fast and stable shape function computations.

We will consider finite elements on axis-parallel hexahedra (bricks) only. Other types including quadrilateral, triangular, prismatic and tetrahedral elements as well as $\mathbf{H}(\mathrm{div})$-conforming elements can also be constructed, see e.g. [41, Chapter 5].

2.2.1 Orthogonal polynomials

As building blocks for shape functions we use orthogonal polynomials, namely the *Legendre polynomials.* Their short definition and some necessary properties will be given, while we refer to books e.g. [36] for further details. Let us define the polynomials on interval $[-1, 1]$ recursively

$$\begin{aligned} l_0(x) &= 1, \\ l_1(x) &= x, \\ l_{j+1}(x) &= \frac{1}{j+1}((2j+1)l_j(x)x - jl_{j-1}(x)), \quad j \in \mathbb{N}. \end{aligned} \tag{2.1}$$

The Legendre polynomials $\{l_j\}_{0\le j\le p}$ form a $L^2([-1,1])$-orthogonal family spanning $\mathcal{P}^p([-1,1])$

$$\int_{-1}^{1} l_i(x)l_j(x)\,d\mathbf{x} = \frac{2}{2j+1}\delta_{ij}.$$

In fact we are going to use the integrated Legendre polynomials. They are defined as

$$L_j(x) = \int_{-1}^{x} l_{j-1}(y)dy \qquad \text{for } x\in[-1,1] \text{ and } j\ge 2.$$

As well as the Legendre polynomials they also can be defined recursively

$$\begin{aligned}
\tilde{L}_1(x) &= x,\\
L_2(x) &= \frac{1}{2}(x^2-1),\\
L_{j+1}(x) &= \frac{1}{j+1}((2j-1)L_j(x)x-(j-2)l_{j-1}(x)), \quad j\ge 2.
\end{aligned} \tag{2.2}$$

Note that here $\tilde{L}_1(x)$ replaces $L_1(x)=x+1$ in order to make the recursive definition working.

The integrated Legendre polynomials $\{L_j\}_{2\le j\le p}$ form an orthogonal family with respect to $H^1([-1,1])$-seminorm

$$\int_{-1}^{1} L_i'(x)L_j'(x)\,dx = 0 \quad \text{for } i\ne j,$$

moreover, they are "almost" L^2-orthogonal

$$\int_{-1}^{1} L_i(x)L_j(x)\,dx = 0 \quad \text{for } |i-j|>2.$$

For $j\ge 2$ the polynomials vanish at the boundary points, $L_j(-1)=L_j(1)=0$, so $\{L_j\}_{2\le j\le p}$ span $\mathcal{P}_0^p[-1,1]$.

2.2.2 Element-based spatial variables

For simplicity we consider the hexahedral element only. First we look at the reference cell and introduce a local numbering of vertices, edges and faces. The cell has 8 vertices $v_1,\dots,v_8$ with the coordinates

$$\begin{aligned}
\mathbf{z}_{v_1} &= (0,0,0), & \mathbf{z}_{v_5} &= (0,0,1),\\
\mathbf{z}_{v_2} &= (1,0,0), & \mathbf{z}_{v_6} &= (1,0,1),\\
\mathbf{z}_{v_3} &= (1,1,0), & \mathbf{z}_{v_7} &= (1,1,1),\\
\mathbf{z}_{v_4} &= (0,1,0), & \mathbf{z}_{v_8} &= (0,1,1),
\end{aligned}$$

there are 12 edges formed by the vertices with numbers

$$\begin{aligned}
e_1 &= (2,1), & e_5 &= (5,1), & e_9 &= (6,5),\\
e_2 &= (3,2), & e_6 &= (6,2), & e_{10} &= (7,6),\\
e_3 &= (4,3), & e_7 &= (7,3), & e_{11} &= (8,7),\\
e_4 &= (4,1), & e_8 &= (8,4), & e_{12} &= (8,5),
\end{aligned}$$

and 6 faces formed by the vertices with numbers

$$\begin{aligned}
f_1 &= (2,1,4,3), & f_4 &= (3,4,8,7),\\
f_2 &= (1,2,6,5), & f_5 &= (4,1,5,8),\\
f_3 &= (2,3,7,6), & f_6 &= (5,6,7,8).
\end{aligned}$$

For every vertex v_j let us define auxiliary functions λ_{v_j} and σ_{v_j}

$$\begin{aligned}
\lambda_{v_1} &= (1-x)(1-y)(1-z), & \sigma_{v_1} &= (1-x)+(1-y)+(1-z),\\
\lambda_{v_2} &= x(1-y)(1-z), & \sigma_{v_2} &= x+(1-y)+(1-z),\\
\lambda_{v_3} &= xy(1-z), & \sigma_{v_3} &= x+y+(1-z),\\
\lambda_{v_4} &= (1-x)y(1-z), & \sigma_{v_4} &= (1-x)+y+(1-z),\\
\lambda_{v_5} &= (1-x)(1-y)z, & \sigma_{v_5} &= (1-x)+(1-y)+z,\\
\lambda_{v_6} &= x(1-y)z, & \sigma_{v_6} &= x+(1-y)+z,\\
\lambda_{v_7} &= xyz, & \sigma_{v_7} &= x+y+z\\
\lambda_{v_8} &= (1-x)yz, & \sigma_{v_8} &= (1-x)+y+z.
\end{aligned}$$

λ_{v_j} is equal to one at the associated vertex and zero at others. Now we define another handy functions based on λ_{v_j} and σ_{v_j}.

- For an edge $e_i = (v_1, v_2)$ there are a parametrization ξ_{e_i}

  $$\xi_{e_i} = \sigma_{v_2} - \sigma_{v_1} \qquad \text{s.t. } \xi_{e_i} : \mathbb{R}^3 \supset [\mathbf{z}_{v_1}, \mathbf{z}_{v_2}] \to [-1, 1]$$

  and the edge extension parameter $\lambda_{e_i} = \lambda_{v_1} + \lambda_{v_2}$ s.t. it is equal to one on the edge e_i and zero on edges parallel to e_i. Moreover, the tangential vector for the edge is given by a simple formula $\tau_{e_i} = \frac{1}{2}\nabla\xi_{e_i}$.

- For a face $f_i = (v_1, v_2, v_3, v_4)$ (v_1 and v_3 are not connected by an edge) there are parametrizations ξ_{f_i} and η_{f_i}

  $$\begin{aligned}
  \xi_{f_i} &= \sigma_{v_2} - \sigma_{v_1} & &\text{s.t. } \xi_{f_i} : \mathbb{R}^3 \supset [\mathbf{z}_{v_1}, \mathbf{z}_{v_2}] \to [-1, 1],\\
  \eta_{f_i} &= \sigma_{v_4} - \sigma_{v_1} & &\text{s.t. } \eta_{f_i} : \mathbb{R}^3 \supset [\mathbf{z}_{v_1}, \mathbf{z}_{v_4}] \to [-1, 1]
  \end{aligned}$$

  and the face extension parameter $\lambda_{f_i} = \lambda_{v_1} + \lambda_{v_2} + \lambda_{v_3} + \lambda_{v_4}$ s.t. it is equal to one on the face f_i and zero on the opposite face. Moreover, the tangential vectors for the face is given by $\tau_{f_i,\xi} = \frac{1}{2}\nabla\xi_{f_i}$, $\tau_{f_i,\eta} = \frac{1}{2}\nabla\eta_{f_i}$ and the outer normal by $\mathbf{n}_{f_i} = \nabla\lambda_{f_i}$.

2.2.3 H^1-conforming elements

For construction of high order finite elements it is convenient to divide all shape functions and degrees of freedom into so-called vertex-, edge-, face- and cell-based (*V*-*E*-*F*-*C*) classes. This grouping is very natural because we construct them to hold certain properties on the corresponding elements of mesh.

A H^1-conforming V-E-F-C basis is given by the following construction.

- **The lowest order shape functions (vertex-based)**
  for $i = 1, \dots, 8$ consider the vertex v_i

  $$\phi_{v_i} = \lambda_{v_i}.$$

- **Edge-based shape functions**
  for $i = 1, \dots, 12$ consider the edge $e_i = (v_1, v_2)$,
  for $l = 0, \dots, p_{e_i} - 2$

  $$\phi^l_{e_i} = L_{l+2}(-\xi_{e_i})\lambda_{e_i},$$

  where $\xi_{e_i} = \sigma_{v_2} - \sigma_{v_1}$ and $\lambda_{e_i} = \lambda_{v_1} + \lambda_{v_2}$.

- **Face-based shape functions**
  for $i = 1, \dots, 6$ consider the face $f_i = (v_1, v_2, v_3, v_4)$,
  for $l, m = 0, \dots, p_{f_i} - 2$

$$\phi_{f_i}^{l,m} = L_{l+2}(\xi_{f_i}) L_{m+2}(\eta_{f_i}) \lambda_{f_i},$$

  where $\xi_{f_i} = \sigma_{v_2} - \sigma_{v_1}$, $\eta_{f_i} = \sigma_{v_4} - \sigma_{v_1}$ and $\lambda_{f_i} = \lambda_{v_1} + \lambda_{v_2} + \lambda_{v_3} + \lambda_{v_4}$.

- **Cell-based shape functions**
  for $l, m, n = 0, \dots, p_c - 2$

$$\phi_c^{l,m,n} = L_{l+2}(2x-1) L_{m+2}(2y-1) L_{n+2}(2z-1).$$

As Q_V, Q_E, Q_F, Q_C we denote the local subspaces spanned by the corresponding vertex-, edge-, face- and cell-based shape functions, respectively. So, the local space Q has a decomposition

$$Q(\hat{\Omega}) = Q_V(\hat{\Omega}) \oplus Q_E(\hat{\Omega}) \oplus Q_F(\hat{\Omega}) \oplus Q_C(\hat{\Omega}).$$

Lemma 2.2. *(we refer to [41, Theorem 5.11])*
The shape functions above are linearly independent and H^1-conforming. For a uniform polynomial order $p = p_{e_i} = p_{f_i} = p_c$ they form a basis of $\mathcal{Q}^{p,p,p}(\hat{\Omega})$.

2.2.4 H(curl)-conforming elements

A **H**(curl)-conforming E-F-C basis which explicitly includes gradients of the H^1-conforming basis is given by the following construction.

- **The lowest order shape functions (edge-based)**
  for $i = 1, \dots, 12$ consider the edge $e_i = (v_1, v_2)$

$$\psi_{e_i} = -\frac{1}{2} \lambda_{e_i} \nabla \xi_{e_i},$$

  where $\xi_{e_i} = \sigma_{v_2} - \sigma_{v_1}$ and $\lambda_{e_i} = \lambda_{v_1} + \lambda_{v_2}$.

- **Edge-based shape functions**
  for $i = 1, \dots, 12$ consider the edge $e_i = (v_1, v_2)$,
  for $l = 0, \dots, p_{e_i} - 1$

$$\psi_{e_i}^l = \nabla \phi_{e_i}^l = \nabla \left(L_{l+2}(-\xi_{e_i}) \lambda_{e_i} \right).$$

- **Face-based shape functions**
  for $i = 1, \dots, 6$ consider the face $f_i = (v_1, v_2, v_3, v_4)$,
  for $l, m = 0, \dots, p_{f_i} - 1$
  Type 1

$$\psi_{1,f_i}^{l,m} = \nabla \phi_{f_i}^{l,m} = \nabla \left(L_{l+2}(\xi_{f_i}) L_{m+2}(\eta_{f_i}) \lambda_{f_i} \right),$$

  Type 2

$$\psi_{2,f_i}^{l,m} = \lambda_{f_i} \left(L'_{l+2}(\xi_{f_i}) L_{m+2}(\eta_{f_i}) \nabla \xi_{f_i} - L_{l+2}(\xi_{f_i}) L'_{m+2}(\eta_{f_i}) \nabla \eta_{f_i} \right),$$

  Type 3

$$\psi_{3,f_i}^{*,m} = \lambda_{f_i} L_{m+2}(\eta_{f_i}) \nabla \xi_{f_i},$$
$$\psi_{3,f_i}^{l,*} = \lambda_{f_i} L_{l+2}(\xi_{f_i}) \nabla \eta_{f_i},$$

  where $\xi_{f_i} = \sigma_{v_2} - \sigma_{v_1}$, $\eta_{f_i} = \sigma_{v_4} - \sigma_{v_1}$ and $\lambda_{f_i} = \lambda_{v_1} + \lambda_{v_2} + \lambda_{v_3} + \lambda_{v_4}$.

- **Cell-based shape functions**
  for $l, m, n = 0, \dots, p_c - 1$
  Type 1
$$\psi_{1,c}^{l,m,n} = \nabla \phi_c^{l,m,n} = \nabla \left(L_{l+2}(2x-1) L_{m+2}(2y-1) L_{n+2}(2z-1) \right),$$
  Type 2
$$\psi_{21,c}^{l,m,n} = \mathrm{diag}\{1,-1,1\} \psi_{1,c}^{l,m,n},$$
$$\psi_{22,c}^{l,m,n} = \mathrm{diag}\{1,1,-1\} \psi_{1,c}^{l,m,n},$$
  Type 3
$$\psi_{3,c}^{*,m,n} = L_{m+2}(2y-1) L_{n+2}(2z-1) \mathbf{e}_x,$$
$$\psi_{3,c}^{l,*,n} = L_{l+2}(2x-1) L_{n+2}(2z-1) \mathbf{e}_y,$$
$$\psi_{3,c}^{l,m,*} = L_{l+2}(2x-1) L_{m+2}(2y-1) \mathbf{e}_z.$$

This basis may also be called a N-E-F-C basis, where N means the lowest order Nédélec elements.

As $\mathbf{X}_N, \mathbf{X}_E, \mathbf{X}_F, \mathbf{X}_C$ we denote the local subspaces spanned by the corresponding lowest-order, edge-, face- and cell-based shape functions, respectively. The local space $\mathbf{X}$ can be represented as
$$\mathbf{X}(\hat{\Omega}) = \mathbf{X}_N(\hat{\Omega}) \oplus \mathbf{X}_E(\hat{\Omega}) \oplus \mathbf{X}_F(\hat{\Omega}) \oplus \mathbf{X}_C(\hat{\Omega}).$$

Lemma 2.3. *(we refer to [41, Theorem 5.12])*
The shape functions above are linearly independent and $\mathbf{H}(\mathrm{curl})$*-conforming. For a uniform polynomial order* $p = p_{e_i} = p_{f_i} = p_c$ *the shape functions form a basis of* $\mathcal{Q}^{p,p+1,p+1}(\hat{\Omega}) \times \mathcal{Q}^{p+1,p,p+1}(\hat{\Omega}) \times \mathcal{Q}^{p+1,p+1,p}(\hat{\Omega})$.

As we have seen, all the shape functions can be defined in terms of the functions L_m, ξ, η, λ and their gradients. Moreover, since the functions $\xi_{e_i}, \lambda_{e_i}, \xi_{f_i}, \eta_{f_i}, \lambda_{f_i}$ are just linear combinations of $\{\lambda_{v_i}\}$ and $\{\sigma_{v_i}\}$ one needs to define only these 16 functions and their gradients. L_m and $L'_m = l_{m-1}$ can be computed recursively according to (2.1) and (2.2). As a result we have a very general and implementation-friendly construction.

Assembling of the global high order finite element spaces is done via the reference element and mapping, in the same way as it was described for the lowest order finite elements. A tricky point is the orientation problem. The global basis functions are constructed from shape functions on neighbor elements, which have a common edge or face. A shape function related to the common edge or face could jump on boundary of the neighbor elements. It happens because the shape function is defined in a coordinate system of an element and the systems may differ on the neighbor elements. For the high order finite elements, which may have multiple kinds of shape functions associated with a face or edge, it is also a problem to join the right shape functions. Our treatment of the orientation problem is implementation-specific and will be explained later in Chapter 5.

Lemma 2.4. *(we refer to [41, Theorem 5.32])*
The high order finite elements described above satisfy the exact sequence property on a finer level, namely they have the local exact sequence property. According to the V-E-F-C (N-E-F-C) structure the global finite element spaces Q_h *and* $\mathbf{X}_h$ *can be represented as*
$$Q_h = Q_{h,V} \oplus \sum_{e \in \mathcal{E}_h} Q_{E_e} \oplus \sum_{f \in \mathcal{F}_h} Q_{F_f} \oplus \sum_{c \in \mathcal{C}_h} Q_{C_c},$$
$$\mathbf{X}_h = \mathbf{X}_{h,N} \oplus \sum_{e \in \mathcal{E}_h} \mathbf{X}_{E_e} \oplus \sum_{f \in \mathcal{F}_h} \mathbf{X}_{F_f} \oplus \sum_{c \in \mathcal{C}_h} \mathbf{X}_{C_c}.$$

Their subspaces form the following exact sequences (the range of each discrete differential operator coincides with the kernel of the following operator)

$$Q_{h,V} \xrightarrow{\nabla} \mathbf{X}_{h,N}, \quad Q_{E_e} \xrightarrow{\nabla} \mathbf{X}_{E_e} (\textit{bijective}), \quad Q_{F_f} \xrightarrow{\nabla} \mathbf{X}_{F_f}, \quad Q_{C_c} \xrightarrow{\nabla} \mathbf{X}_{C_c},$$

for all $e \in \mathcal{E}_h$, $f \in \mathcal{F}_h$, $c \in \mathcal{C}_h$.

2.2.5 H^1-conforming degrees of freedom

The original definition of the degrees of freedom given in [29] and [30] was made for a uniform polynomial order and does not allow varying polynomial order. For the latter case one needs another degrees of freedom, e.g. the ones presented in [11].

Below we use the following notations, integrals of kind $\int_f$ denote integration over a face $f \in \mathcal{F}_h$ (edge e or cell c), spaces of kind $\mathcal{P}^{p_e}(e)$ or $\mathcal{Q}_0^{p_f,p_f}(f)$ denote the polynomial spaces over an edge e or face f, the index 0 means a subspace which vanishes on the boundary.

The following degrees of freedom define an *unisolvent* H^1-conforming finite elements.

- **Vertex-based degrees of freedom**
  for a vertex $v \in \mathcal{V}_h$ with the coordinates $\mathbf{z}_v$,
$$\ell_v(q) = q(\mathbf{z}_v).$$
- **Edge-based degrees of freedom**
  for an edge $e \in \mathcal{E}_h$ with the element of edge-order p_e,
$$\ell_e^l(q) = \int_e \frac{\partial q}{\partial s}\frac{\partial \mathsf{p}_l}{\partial s} ds \qquad \text{for } l = 0, \dots, p_e - 2,$$
  where $\{\mathsf{p}_l\}$ is a basis of $\mathcal{P}_0^{p_e}(e)$.
- **Face-based degrees of freedom**
  for a face $f \in \mathcal{F}_h$ with the element of face-order p_f,
$$\ell_f^{l,m}(q) = \int_f \nabla_f q \cdot \nabla_f \mathsf{p}_{l,m} \, dA \qquad \text{for } l, m = 0, \dots, p_f - 2,$$
  where $\{\mathsf{p}_{l,m}\}$ is a basis of $\mathcal{Q}_0^{p_f,p_f}(f)$ and the surface gradient ∇_f is defined as $\nabla_f q = \mathbf{n}_f \times \nabla q \times \mathbf{n}_f$.
- **Cell-based degrees of freedom**
  for a cell $c \in \mathcal{C}_h$ with the element of cell-order p_c,
$$\ell_c^{l,m,n}(q) = \int_c \nabla q \cdot \nabla \mathsf{p}_{l,m,n} \, d\mathbf{x} \qquad \text{for } l, m, n = 0, \dots, p_c - 2,$$
  where $\{\mathsf{p}_{l,m,n}\}$ is a basis of $\mathcal{Q}_0^{p_c,p_c,p_c}(c)$.

2.2.6 H(curl)-conforming degrees of freedom

The following degrees of freedom define a unisolvent **H**(curl)-conforming finite elements.

- **Edge-based degrees of freedom**
  for an edge $e \in \mathcal{E}_h$ with the element of edge-order p_e,
$$\ell_e^l(\mathbf{u}) = \int_e (\mathbf{u} \cdot \mathbf{t}_e) \mathsf{p}_l \, ds \qquad \text{for } l = 0, \dots, p_e,$$
  where $\{\mathsf{p}_l\}$ is a basis of $\mathcal{P}^{p_e}(e)$ and $\mathbf{t}_e$ is the unit tangent vector.

- **Face-based degrees of freedom**
  for a face $f \in \mathcal{F}_h$ with the element of face-order p_f,

$$\ell_f^l(\mathbf{u}) = \int_f \mathrm{curl}_f\, \mathbf{u} \cdot \mathbf{v}_l \, dA \qquad \text{for } \{\mathbf{v}_l\} \text{ a basis of } \mathcal{Q}_{\mathrm{curl}_f},$$

$$\ell_f^l(\mathbf{u}) = \int_f \mathbf{u} \cdot \mathbf{v}_l \, dA \qquad \text{for } \{\mathbf{v}_l\} \text{ a basis of } \mathcal{Q}_{\nabla_f},$$

  where the surface rotor curl_f is defined as $\mathrm{curl}_f\, \mathbf{v} = (\nabla \times \mathbf{v}) \cdot \mathbf{n}_f$ and

$$\mathcal{Q}_{\mathrm{curl}_f} = \{\mathrm{curl}_f\, \mathbf{v} \mid \mathbf{v} \in (\mathcal{Q}^{p_f, p_f+1}(f) \times \mathcal{Q}^{p_f+1, p_f}(f)) \cap \mathbf{H}_0(\mathrm{curl}, f)\},$$
$$\mathcal{Q}_{\nabla_f} = \{\nabla_f\, p \mid p \in \mathcal{Q}_0^{p_f+1, p_f+1}(f)\}.$$

- **Cell-based degrees of freedom**
  for a cell $c \in \mathcal{C}_h$ with the element of cell-order p_c,

$$\ell_c^l(\mathbf{u}) = \int_c \nabla \times \mathbf{u} \cdot \mathbf{v}_l \, d\mathbf{x} \qquad \text{for } \{\mathbf{v}_l\} \text{ a basis of } \mathcal{Q}_{\mathrm{curl}},$$

$$\ell_c^l(\mathbf{u}) = \int_c \mathbf{u} \cdot \mathbf{v}_l \, d\mathbf{x} \qquad \text{for } \{\mathbf{v}_l\} \text{ a basis of } \mathcal{Q}_{\nabla},$$

  where $p_1 = p_c + 1$ and

$$\mathcal{Q}_{\mathrm{curl}} = \{\nabla \times \mathbf{v} \mid \mathbf{v} \in \mathcal{Q}^{p_c, p_1, p_1} \times \mathcal{Q}^{p_1, p_c, p_1} \times \mathcal{Q}^{p_1, p_1, p_c}(c) \cap \mathbf{H}_0(\mathrm{curl}, c)\},$$
$$\mathcal{Q}_{\nabla} = \{\nabla p \mid p \in \mathcal{Q}_0^{p_c+1, p_c+1, p_c+1}(c)\}.$$

As we have seen, the degrees of freedom are rather complex. In case one uses the finite elements with a uniform polynomial order, it is easier to use the classical degrees of freedom described in [29].

There is some freedom in tuning the degrees of freedom because the test functions p_l and $\mathbf{v}_l$ are given in terms of a basis. One can choose the concrete p_l and $\mathbf{v}_l$ and so fix the degrees of freedom. It may be used to obtain degrees of freedom which are dual to some shape functions. Particularly, it is possible to choose such $\{\mathsf{p}_l\}$ and $\{\mathbf{v}_l\}$ that the shape functions and the corresponding degrees of freedom on the reference element form a dual basis (nodal basis) inside the vertex-, edge-, face- and cell-based groups, e.g. $\ell_e^l(\psi_e^m) = \delta_{lm}$ for the edge-based shape functions and degrees of freedom. This can be done by solving a linear system in the form $\ell^l(\psi^m) = \delta_{lm}$ with respect to a parametrized representation of p_l and $\mathbf{v}_l$.

In our case this task was done in the *Wolfram's Mathematica* by using symbolic calculations. In the first step the shape functions and parametrized degrees of freedom were constructed in Mathematica. Then, the corresponding linear systems $\ell_e^l(\psi_e^m) = \delta_{lm}$ were solved with respect to parameters in the degrees of freedom. The resulted parameters provide the local nodal bases inside the V-E-F-C parts.

2.2.7 Interpolation operators

Let $\{\varphi_j\}$ be a basis of the high order finite element space Q_h. Our goal is to construct an operator Π_{Q_h} which maps a sufficiently smooth $q \in H^1_{\mathrm{per}}(\Omega)$ to Q_h by providing coefficients a_j in the basis, s.t.

$$\Pi_{Q_h}(q) = \sum_j a_j \varphi_j.$$

In Section 2.1 the interpolation operators $\Pi_{\mathbf{X}_{h,0}}$ and $\Pi_{Q_{h,0}}$ were defined for the lowest order finite elements via a nodal basis. Now we need to define interpolation operators to the high order finite element spaces. It is not so easy anymore because the nodal basis is not directly available in practice. To overcome this problem one should use the following "triangular" trick.

Our global finite element space is sum of finite elements for all cells $c \in \mathcal{C}_h$. So the interpolation may also be done cell by cell in such a way that every cell makes some contribution to the coefficients. Working on cell level we may notice that, by construction, the cell-based shape functions decay on faces, the face-based shape functions decay on edges and the edge-based shape functions decay on vertices. It follows the very important conclusion, if we number all shape functions $\{\phi_j\}$ and all degrees of freedom $\{\ell_i\}$ in the order they were introduced, then the matrix $\theta_{ij} = \ell_i(\phi_j)$ is block-triangular, its blocks correspond to vertex-, edge-, face- and cell-based parts. If the degrees of freedom were chosen such that the V-E-F-C parts have the local nodal bases, then the matrix θ_{ij} is triangular with the unit matrices as diagonal blocks. We may exploit this structure and get the coefficients quite easily.

For simplicity assume that we act on the reference element. Let us number the shape functions, degrees of freedom and coefficients with a double index (t, j), where $t = 1, 2, 3, 4$ means the type: vertex, edge, face or cell. The second index $j = 1, \dots, n_t$ is the number inside a type. For a given function $q \in H^1_{\text{per}}(\Omega)$ the algorithm to get an interpolation (coefficient vector $(a_{t,j})$) in the local finite element basis is the following.

1. $t := 1$, $f_h := 0$.
2. $a_{t,j} = \ell_{t,j}(q - f_h) = \ell_{t,j}(q) - \ell_{t,j}(f_h)$ for $j = 1, \dots, n_t$.
3. $f_h := f_h + \sum_{j=1}^{n_t} a_{t,j}\phi_{t,j}$.
4. while $t \neq 4$, $t := t + 1$ and go to 2.

In the same way we define the interpolation operator $\Pi_{\mathbf{X}_h}$, which maps to the high order $\mathbf{H}(\text{curl})$-conforming finite element space $\mathbf{X}_h$.

2.2.8 Static condensation

After assembling of the global finite element spaces the cell-based degrees of freedom of a cell are decoupled from ones of other cells. It happens because support of the global cell-based shape functions is restricted within their own cells only.

This fact can be used to decrease the size of a problem, e.g. a linear system. Let us denote quantities related to the cell-based degrees of freedom with the index C and others with the index R. After a renumbering a linear system derived from our finite element discretization may be written in the block form

$$\begin{pmatrix} A_{RR} & A_{RC} \\ A_{CR} & A_{CC} \end{pmatrix} \begin{pmatrix} \mathbf{u}_R \\ \mathbf{u}_C \end{pmatrix} = \begin{pmatrix} \mathbf{f}_R \\ \mathbf{f}_C \end{pmatrix}.$$

Since the cell-based degrees of freedom for different cells are decoupled, the matrix $A_{CC} = \text{diag}(A_{CC}^{c_1}, \dots, A_{CC}^{c_n})$, where $c_1, \dots, c_n \in \mathcal{C}_h$ and $\{A_{CC}^{c_j}\}$ are the element-level matrices. Now we can compute the *Schur complement* with respect to the C-part. It gives the smaller condensed system $A_z\mathbf{u}_R = \mathbf{f}_z$, where

$$A_z = A_{RR} - A_{RC}A_{CC}^{-1}A_{CR},$$
$$\mathbf{f}_z = \mathbf{f}_R - A_{RC}A_{CC}^{-1}\mathbf{f}_C.$$

After the condensed system was solved the cell-based unknowns can be obtained as

$$\mathbf{u}_C = A_{CC}^{-1}(\mathbf{f}_C - A_{CR}\mathbf{u}_R).$$

The static condensation is an important tool for finite elements of order $p \geq 3$ when the fraction of the cell-based degrees of freedom becomes significant. The condensation can be realized on element level with no expensive operations. As a result one gets a smaller and better conditioned system what is advantageous for iterative methods. The condition number is better because the Schur complement means an orthogonalization of the cell-based basis functions with respect to the other ones.

2.2.9 Example of shape functions

The definition of hierarchical high order finite elements is rather complex. As an illustration we provide the exact form of the second order H^1- and $\mathbf{H}(\mathrm{curl})$-conforming elements on hexahedra. The shape functions are defined on the reference element $\hat{\Omega} = [0,1]^3$ according to the local numbering and notations described in Subsections 2.2.2, 2.2.3 and 2.2.4.

H^1-conforming element, total 27 shape functions:

1. The lowest order shape functions (vertex-based)

$$\begin{aligned}
\phi_{v_1} &= (1-x)(1-y)(1-z), & \phi_{v_5} &= (1-x)(1-y)z,\\
\phi_{v_2} &= x(1-y)(1-z), & \phi_{v_6} &= x(1-y)z,\\
\phi_{v_3} &= xy(1-z), & \phi_{v_7} &= xyz,\\
\phi_{v_4} &= (1-x)y(1-z), & \phi_{v_8} &= (1-x)yz.
\end{aligned}$$

2. Edge-based shape functions

$$\begin{aligned}
\phi^0_{e_1} &= 2(x-1)x(y-1)(z-1), & \phi^0_{e_7} &= 2xy(z-1)z,\\
\phi^0_{e_2} &= -2x(y-1)y(z-1), & \phi^0_{e_8} &= -2(x-1)y(z-1)z,\\
\phi^0_{e_3} &= -2(x-1)xy(z-1), & \phi^0_{e_9} &= -2(x-1)x(y-1)z,\\
\phi^0_{e_4} &= 2(x-1)(y-1)y(z-1), & \phi^0_{e_{10}} &= 2x(y-1)yz,\\
\phi^0_{e_5} &= 2(x-1)(y-1)(z-1)z, & \phi^0_{e_{11}} &= 2(x-1)xyz,\\
\phi^0_{e_6} &= -2x(y-1)(z-1)z, & \phi^0_{e_{12}} &= -2(x-1)(y-1)yz.
\end{aligned}$$

3. Face-based shape functions

$$\begin{aligned}
\phi^{0,0}_{f_1} &= -4(x-1)x(y-1)y(z-1), & \phi^{0,0}_{f_4} &= 4(x-1)xy(z-1)z,\\
\phi^{0,0}_{f_2} &= -4(x-1)x(y-1)(z-1)z, & \phi^{0,0}_{f_5} &= -4(x-1)(y-1)y(z-1)z,\\
\phi^{0,0}_{f_3} &= 4x(y-1)y(z-1)z, & \phi^{0,0}_{f_6} &= 4(x-1)x(y-1)yz.
\end{aligned}$$

4. Cell-based shape function

$$\phi^{0,0,0}_c = 8(x-1)x(y-1)y(z-1)z.$$

$\mathbf{H}(\mathrm{curl})$-conforming element, total 54 shape functions:

1. The lowest order shape functions (edge-based)

$$\begin{aligned}
\psi_{e_1} &= \{(y-1)(z-1),0,0\}, & \psi_{e_7} &= \{0,0,xy\},\\
\psi_{e_2} &= \{0,-x(z-1),0\}, & \psi_{e_8} &= \{0,0,-(x-1)y\},\\
\psi_{e_3} &= \{y(z-1),0,0\}, & \psi_{e_9} &= \{-(y-1)z,0,0\},\\
\psi_{e_4} &= \{0,(x-1)(z-1),0\}, & \psi_{e_{10}} &= \{0,xz,0\},\\
\psi_{e_5} &= \{0,0,(x-1)(y-1)\}, & \psi_{e_{11}} &= \{yz,0,0\},\\
\psi_{e_6} &= \{0,0,-x(y-1)\}, & \psi_{e_{12}} &= \{0,-(x-1)z,0\}.
\end{aligned}$$

2. Edge-based shape functions

$$\begin{aligned}
\psi^0_{e_1} &= \{2(2x-1)(y-1)(z-1), 2(x-1)x(z-1), 2(x-1)x(y-1)\},\\
\psi^0_{e_2} &= \{-2(y-1)y(z-1), -2x(2y-1)(z-1), -2x(y-1)y\},\\
\psi^0_{e_3} &= \{-2(2x-1)y(z-1), -2(x-1)x(z-1), -2(x-1)xy\},\\
\psi^0_{e_4} &= \{2(y-1)y(z-1), 2(x-1)(2y-1)(z-1), 2(x-1)(y-1)y\},\\
\psi^0_{e_5} &= \{2(y-1)(z-1)z, 2(x-1)(z-1)z, 2(x-1)(y-1)(2z-1)\},\\
\psi^0_{e_6} &= \{-2(y-1)(z-1)z, -2x(z-1)z, -2x(y-1)(2z-1)\},\\
\psi^0_{e_7} &= \{2y(z-1)z, 2x(z-1)z, 2xy(2z-1)\},\\
\psi^0_{e_8} &= \{-2y(z-1)z, -2(x-1)(z-1)z, -2(x-1)y(2z-1)\},\\
\psi^0_{e_9} &= \{-2(2x-1)(y-1)z, -2(x-1)xz, -2(x-1)x(y-1)\},\\
\psi^0_{e_{10}} &= \{2(y-1)yz, 2x(2y-1)z, 2x(y-1)y\},\\
\psi^0_{e_{11}} &= \{2(2x-1)yz, 2(x-1)xz, 2(x-1)xy\},\\
\psi^0_{e_{12}} &= \{-2(y-1)yz, -2(x-1)(2y-1)z, -2(x-1)(y-1)y\}.
\end{aligned}$$

3. Face-based shape functions

$$\begin{aligned}
\psi^{0,0}_{1,f_1} &= 4\{-(2x-1)(y-1)y(z-1), -(x-1)x(2y-1)(z-1), -(x-1)x(y-1)y\},\\
\psi^{0,0}_{1,f_2} &= 4\{-(2x-1)(y-1)(z-1)z, -(x-1)x(z-1)z, -(x-1)x(y-1)(2z-1)\},\\
\psi^{0,0}_{1,f_3} &= 4\{(y-1)y(z-1)z, x(2y-1)(z-1)z, x(y-1)y(2z-1)\},\\
\psi^{0,0}_{1,f_4} &= 4\{(2x-1)y(z-1)z, (x-1)x(z-1)z, (x-1)xy(2z-1)\},\\
\psi^{0,0}_{1,f_5} &= 4\{-(y-1)y(z-1)z, -(x-1)(2y-1)(z-1)z, -(x-1)(y-1)y(2z-1)\},\\
\psi^{0,0}_{1,f_6} &= 4\{(2x-1)(y-1)yz, (x-1)x(2y-1)z, (x-1)x(y-1)y\},
\end{aligned}$$

$$\begin{aligned}
\psi^{0,0}_{2,f_1} &= \{-4(2x-1)(y-1)y(z-1), 4(x-1)x(2y-1)(z-1), 0\},\\
\psi^{0,0}_{2,f_2} &= \{-4(2x-1)(y-1)(z-1)z, 0, 4(x-1)x(y-1)(2z-1)\},\\
\psi^{0,0}_{2,f_3} &= \{0, 4x(2y-1)(z-1)z, -4x(y-1)y(2z-1)\},\\
\psi^{0,0}_{2,f_4} &= \{4(2x-1)y(z-1)z, 0, -4(x-1)xy(2z-1)\},\\
\psi^{0,0}_{2,f_5} &= \{0, -4(x-1)(2y-1)(z-1)z, 4(x-1)(y-1)y(2z-1)\},\\
\psi^{0,0}_{2,f_6} &= \{4(2x-1)(y-1)yz, -4(x-1)x(2y-1)z, 0\},
\end{aligned}$$

$$\begin{aligned}
\psi^{*,0}_{3,f_1} &= \{4(y-1)y(z-1), 0, 0\}, & \psi^{0,*}_{3,f_1} &= \{0, -4(x-1)x(z-1), 0\},\\
\psi^{*,0}_{3,f_2} &= \{-4(y-1)(z-1)z, 0, 0\}, & \psi^{0,*}_{3,f_2} &= \{0, 0, -4(x-1)x(y-1)\},\\
\psi^{*,0}_{3,f_3} &= \{0, 4x(z-1)z, 0\}, & \psi^{0,*}_{3,f_3} &= \{0, 0, 4x(y-1)y\},\\
\psi^{*,0}_{3,f_4} &= \{-4y(z-1)z, 0, 0\}, & \psi^{0,*}_{3,f_4} &= \{0, 0, 4(x-1)xy\},\\
\psi^{*,0}_{3,f_5} &= \{0, 4(x-1)(z-1)z, 0\}, & \psi^{0,*}_{3,f_5} &= \{0, 0, -4(x-1)(y-1)y\},\\
\psi^{*,0}_{3,f_6} &= \{4(y-1)yz, 0, 0\}, & \psi^{0,*}_{3,f_6} &= \{0, 4(x-1)xz, 0\}.
\end{aligned}$$

4. Cell-based shape functions

$$\begin{aligned}
\psi^{0,0,0}_{1,c} &= 8\{(2x-1)(y-1)y(z-1)z, (x-1)x(2y-1)(z-1)z, (x-1)x(y-1)y(2z-1)\},\\
\psi^{0,0,0}_{21,c} &= 8\{(2x-1)(y-1)y(z-1)z, -(x-1)x(2y-1)(z-1)z, (x-1)x(y-1)y(2z-1)\},\\
\psi^{0,0,0}_{22,c} &= 8\{(2x-1)(y-1)y(z-1)z, (x-1)x(2y-1)(z-1)z, -(x-1)x(y-1)y(2z-1)\},\\
\psi^{*,0,0}_{3,c} &= \{4(y-1)y(z-1)z, 0, 0\},\\
\psi^{0,*,0}_{3,c} &= \{0, 4(x-1)x(z-1)z, 0\},\\
\psi^{0,0,*}_{3,c} &= \{0, 0, 4(x-1)x(y-1)y\}.
\end{aligned}$$

2.3 Modified elements

Following [13] and [8] we introduce k-modified finite elements. We will consider the lowest order elements first and then indicate changes for the high order elements.

The nodal bases $\{\psi_{e,\mathbf{0}}\}$ and $\{\phi_{v,\mathbf{0}}\}$ of the standard lowest order elements defined in Section 2.1 are modified by the multiplication by an exponential factor

$$\begin{aligned} \psi_{e,\mathbf{k}}(\mathbf{x}) &= \mathrm{e}^{-\mathrm{i}\mathbf{k}\cdot(\mathbf{x}-\mathbf{m}_e)}\psi_{e,\mathbf{0}}(\mathbf{x}) \qquad e \in \mathcal{E}_h, \\ \phi_{v,\mathbf{k}}(\mathbf{x}) &= \mathrm{e}^{-\mathrm{i}\mathbf{k}\cdot(\mathbf{x}-\mathbf{z}_v)}\phi_{v,\mathbf{0}}(\mathbf{x}) \qquad v \in \mathcal{V}_h. \end{aligned}$$

In practice it can be attained by multiplying the corresponding shape functions during assembling of the global finite element spaces. Note that every basis function gets its own shift $\mathbf{m}_e$ or $\mathbf{z}_v$ depending on the nodal point. The modified bases form new finite elements

$$\begin{aligned} \mathbf{X}_{h,\mathbf{k}} &= \mathrm{span}\{\psi_{e,\mathbf{k}} \colon e \in \mathcal{E}_h\}, \\ Q_{h,\mathbf{k}} &= \mathrm{span}\{\phi_{v,\mathbf{k}} \colon v \in \mathcal{V}_h\}. \end{aligned}$$

To keep the duality the inverse exponential factor is used for the degrees of freedom

$$\ell_{e,\mathbf{k}}(\mathbf{u}) = \int_{\mathbf{x}_e}^{\mathbf{y}_e} \mathrm{e}^{\mathrm{i}\mathbf{k}\cdot(\mathbf{x}-\mathbf{m}_e)}\mathbf{u}\cdot\mathbf{t}_e ds, \tag{2.3}$$

$$\ell_{v,\mathbf{k}}(q) = \left(\mathrm{e}^{\mathrm{i}\mathbf{k}\cdot(\mathbf{x}-\mathbf{z}_v)}q(\mathbf{x})\right)|_{\mathbf{x}=\mathbf{z}_v} = q(\mathbf{z}_v) = \ell_{v,\mathbf{0}}(q). \tag{2.4}$$

Again, the shift is different for every degree of freedom. From (2.4) one may notice that due to our choice of the shift the degrees of freedom for $Q_{h,\mathbf{k}}$ are actually not changed. Since the exponential factor is shifted with respect to the nodal point, periodic fields $q \in H^1_{\mathrm{per}}(\Omega)$ and $\mathbf{u} \in \mathbf{H}_{\mathrm{per}}(\mathrm{curl}, \Omega)$ have the same degrees of freedom on the identified elements of the periodic boundary.

By the construction there exists a nice property connecting the standard and modified elements

$$\begin{aligned} \ell_{e,\mathbf{k}}(\psi_{e,\mathbf{k}}) &= \int_{\mathbf{x}_e}^{\mathbf{y}_e} \mathrm{e}^{\mathrm{i}\mathbf{k}\cdot(\mathbf{x}-\mathbf{m}_e)}\mathrm{e}^{-\mathrm{i}\mathbf{k}\cdot(\mathbf{x}-\mathbf{m}_e)}\psi_{e,\mathbf{0}}(\mathbf{x})\cdot\mathbf{t}_e ds = \ell_{e,\mathbf{0}}(\psi_{e,\mathbf{0}}) = 1, \\ \ell_{v,\mathbf{k}}(\phi_{v,\mathbf{k}}) &= \mathrm{e}^{-\mathrm{i}\mathbf{k}\cdot(\mathbf{x}-\mathbf{z}_v)}\phi_{v,\mathbf{0}}(\mathbf{x})|_{\mathbf{x}=\mathbf{z}_v} = \phi_{v,\mathbf{0}}(\mathbf{z}_v) = \ell_{v,\mathbf{0}}(\phi_{v,\mathbf{0}}) = 1. \end{aligned} \tag{2.5}$$

Interpolation operators to the modified elements are defined as usual

$$\begin{aligned} \Pi_{\mathbf{X}_{h,\mathbf{k}}}(\mathbf{u}) &= \sum_{e\in\mathcal{E}_h} \ell_{e,\mathbf{k}}(\mathbf{u})\psi_{e,\mathbf{k}}, \\ \Pi_{Q_{h,\mathbf{k}}}(q) &= \sum_{v\in\mathcal{V}_h} \ell_{v,\mathbf{k}}(q)\phi_{v,\mathbf{k}}. \end{aligned}$$

Doing straightforward computations one can easily check that the bases $\{\psi_{e,\mathbf{k}}\}$ and $\{\phi_{v,\mathbf{k}}\}$ have the following properties

$$\begin{aligned} \nabla_{\mathbf{k}}\, \phi_{v,\mathbf{k}}(\mathbf{x}) &= \mathrm{e}^{-\mathrm{i}\mathbf{k}\cdot(\mathbf{x}-\mathbf{z}_v)}\nabla\phi_{v,\mathbf{0}}(\mathbf{x}), \\ \nabla_{\mathbf{k}} \times \psi_{e,\mathbf{k}}(\mathbf{x}) &= \mathrm{e}^{-\mathrm{i}\mathbf{k}\cdot(\mathbf{x}-\mathbf{m}_e)}\nabla \times \psi_{e,\mathbf{0}}(\mathbf{x}). \end{aligned} \tag{2.6}$$

The approach presented above can be generalized to high order finite elements. One needs to modify the shape functions and degrees of freedom in the same way as it was done for the lowest order elements. So we put the multiplier $\mathrm{e}^{-\mathrm{i}\mathbf{k}\cdot(\mathbf{x}-\mathbf{s})}$ in front of the shape functions and put the inverse multiplier $\mathrm{e}^{\mathrm{i}\mathbf{k}\cdot(\mathbf{x}-\mathbf{s})}$ under the integrals for the degrees of freedom. Since the high order elements have a more complex V-E-F-C structure, the exponential multiplier $\mathrm{e}^{-\mathrm{i}\mathbf{k}\cdot(\mathbf{x}-\mathbf{s})}$ should have a different shift $\mathbf{s}$ for vertex-, edge-, face- and cell-based basis functions and degrees of freedom. The shift has to be a point on the corresponding element of the mesh, e.g. the midpoints of faces. It guarantees that periodic fields $q \in H^1_{\mathrm{per}}(\Omega)$ and $\mathbf{u} \in \mathbf{H}_{\mathrm{per}}(\mathrm{curl}, \Omega)$ have the same degrees of freedom on the identified elements of the periodic boundary. Remember that there could be many basis functions corresponding to one element of the mesh. They can have the same shift or different shifts inside the element, both are correct. For the shift we use the nodal point which is unique for every basis function.

The high order modified elements hold the property (2.6) and a property similar to (2.5). If we denote sets of the non-modified high order shape functions and degrees of freedom as $\{\psi_{j,\mathbf{0}}\}$ and $\{\ell_{j,\mathbf{0}}\}$, then they are connected with the modified sets as

$$\ell_{j,\mathbf{k}}(\psi_{j,\mathbf{k}}) = \ell_{j,\mathbf{0}}(\psi_{j,\mathbf{0}}) \qquad \text{for all } j.$$

In the thesis we use the same notations $\mathbf{X}_{h,\mathbf{k}}$ and $Q_{h,\mathbf{k}}$ for the modified high order elements. If in a certain place it is not indicated which type we consider then it means that it does not matter, otherwise it is given a separate explanation.

The property (2.6) is a key which makes it possible to prove the next lemmas for both the lowest and high order elements.

Lemma 2.5. *(see [8, Lemma 4])*
Let $\mathbf{k} \neq \mathbf{0}$, the spaces $Q_{h,\mathbf{k}}$ and $\mathbf{X}_{h,\mathbf{k}}$ satisfy the following commuting diagram (functions must be chosen such that the interpolation operators make sense)

$$\begin{array}{ccc} H^1_{\mathrm{per}}(\Omega) & \xrightarrow{\nabla_{\mathbf{k}}} & \mathbf{H}_{\mathrm{per}}(\mathrm{curl}, \Omega) \\ \downarrow \Pi_{Q_{h,\mathbf{k}}} & & \downarrow \Pi_{\mathbf{X}_{h,\mathbf{k}}} \\ Q_{h,\mathbf{k}} & \xrightarrow{\nabla_{\mathbf{k}}} & \mathbf{X}_{h,\mathbf{k}}. \end{array}$$

In addition, the spaces and operators above form the exact sequences horizontally.

Lemma 2.6. *Discrete Helmholtz decomposition (see [8, Lemma 7]).*
For given $\mathbf{u}_h \in \mathbf{X}_{h,\mathbf{k}}$, there exist $\mathbf{v}_h \in \mathbf{X}_{h,\mathbf{k}}$ and $q_h \in Q_{h,\mathbf{k}}$ s.t.

$$\begin{aligned} &\mathbf{u}_h = \mathbf{v}_h + \nabla_{\mathbf{k}} q_h, \\ &(\mathbf{v}_h, \nabla_{\mathbf{k}} p_h)_{\mathbf{L}^2} = 0 \qquad \text{for all } p_h \in Q_{h,\mathbf{k}}. \end{aligned}$$

The last line implies that $\mathbf{v}_h \in \mathbf{V}_{h,\mathbf{k}}$.

Lemma 2.7. *(see [8, Lemmas 8 – 10])*
The modified elements do provide ellipticity on $\mathbf{V}_{h,\mathbf{k}}$, weak approximability of $H^1_{\mathrm{per}}(\Omega)$ and strong approximability of $\mathbf{V}_{\mathbf{k}}$.

Combined with Theorems 1.5 – 1.6 it follows that the modified elements can be used to get a spectrally correct approximation of our eigenvalue problem.

Practical efficiency of the modified elements was proved in [12], where they were applied to Problem 1.2.

2.4 Implementation of the modified elements

For solving a problem with the modified elements we need to assemble matrices related to operators of the original problem. The assembling implies a numerical integration of the basis functions. For the modified elements a direct numerical integration is problematic since the shape functions include the exponential multiplier and so are not polynomial anymore. To overcome this difficulty we need to make an important observation about how the sesquilinear forms look on the bases of $\mathbf{X}_{h,\mathbf{k}}$ and $Q_{h,\mathbf{k}}$.

Using the property (2.6) one obtains

$$
\begin{aligned}
a_{\mathbf{k}}(\psi_{e_1,\mathbf{k}}, \psi_{e_2,\mathbf{k}}) &= (\varepsilon^{-1}\nabla_{\mathbf{k}} \times \psi_{e_1,\mathbf{k}}, \nabla_{\mathbf{k}} \times \psi_{e_2,\mathbf{k}})_{\mathbf{L}^2} \\
&= \int_{\Omega} \varepsilon^{-1} \mathrm{e}^{-\mathrm{i}\mathbf{k}\cdot(\mathbf{x}-\mathbf{m}_{e_1})} \nabla \times \psi_{e_1,\mathbf{0}} \cdot \overline{\mathrm{e}^{-\mathrm{i}\mathbf{k}\cdot(\mathbf{x}-\mathbf{m}_{e_2})} \nabla \times \psi_{e_2,\mathbf{0}}}\, d\mathbf{x} \\
&= \mathrm{e}^{\mathrm{i}\mathbf{k}\cdot(\mathbf{m}_{e_1}-\mathbf{m}_{e_2})} (\varepsilon^{-1}\nabla \times \psi_{e_1,\mathbf{0}}, \nabla \times \psi_{e_2,\mathbf{0}})_{\mathbf{L}^2} \\
&= \mathrm{e}^{\mathrm{i}\mathbf{k}\cdot(\mathbf{m}_{e_1}-\mathbf{m}_{e_2})} a_{\mathbf{0}}(\psi_{e_1,\mathbf{0}}, \psi_{e_2,\mathbf{0}}), \\
m(\psi_{e_1,\mathbf{k}}, \psi_{e_2,\mathbf{k}}) &= \mathrm{e}^{\mathrm{i}\mathbf{k}\cdot(\mathbf{m}_{e_1}-\mathbf{m}_{e_2})} m(\psi_{e_1,\mathbf{0}}, \psi_{e_2,\mathbf{0}}), \\
b_{\mathbf{k}}(\psi_{e,\mathbf{k}}, \phi_{v,\mathbf{k}}) &= \mathrm{e}^{\mathrm{i}\mathbf{k}\cdot(\mathbf{m}_{e}-\mathbf{z}_{v})} b_{\mathbf{0}}(\psi_{e,\mathbf{0}}, \phi_{v,\mathbf{0}}), \\
c_{\mathbf{k}}(\phi_{v_1,\mathbf{k}}, \phi_{v_2,\mathbf{k}}) &= \mathrm{e}^{\mathrm{i}\mathbf{k}\cdot(\mathbf{z}_{v_1}-\mathbf{z}_{v_2})} c_{\mathbf{0}}(\phi_{v_1,\mathbf{0}}, \phi_{v_2,\mathbf{0}}).
\end{aligned}
$$

One may notice that the matrices for $\mathbf{k} \neq \mathbf{0}$ are obtained from the matrices for $\mathbf{k} = \mathbf{0}$ by multiplication of their coefficients by a phase shift factor, only the standard shape functions are actually used in the computations.

Although the formulas are given for the lowest order elements the same holds true for the high order ones. To assemble the matrices we go along all pairs of the basis functions $\{\psi_{i,\mathbf{k}}\}$, every function has its own shift $\mathbf{s}_i$, so the exponential factor is $\mathrm{e}^{\mathrm{i}\mathbf{k}\cdot(\mathbf{s}_i-\mathbf{s}_j)}$ for any sesquilinear form $\cdot(\psi_{i,\mathbf{k}}, \psi_{j,\mathbf{k}})$.

Chapter 3

Eigenvalue solver

Although the mixed formulation from Problem 1.2 can be directly used for practical computations, we choose another formulation, which, we believe, has some advantages in the implementation. Note that we always assume $\mathbf{k} \neq \mathbf{0}$, otherwise it is mentioned explicitly.

Problem 3.1. Discrete eigenvalue problem for positive λ_h.
Find a pair of $(\mathbf{u}_h, \lambda_h) \in \mathbf{X}_{h,\mathbf{k}} \times \mathbb{R}^+$, s.t. for all $\mathbf{v}_h \in \mathbf{X}_{h,\mathbf{k}}$

$$a_{\mathbf{k}}(\mathbf{u}_h, \mathbf{v}_h) = \lambda_h \, m(\mathbf{u}_h, \mathbf{v}_h). \tag{3.1}$$

Let us test Problem (3.1) with $\mathbf{v}_h = \nabla_{\mathbf{k}} \, q_h$, where $q_h \in Q_{h,\mathbf{k}}$. Since $\nabla_{\mathbf{k}} \times \nabla_{\mathbf{k}} \, q_h = 0$, it gives

$$0 = \lambda_h \, (\mathbf{u}_h, \nabla_{\mathbf{k}} \, q_h)_{\mathbf{L}^2} = \lambda_h \, b_{\mathbf{k}}(\mathbf{u}_h, q_h). \tag{3.2}$$

From (3.2) it follows that $\mathbf{u}_h \in \mathbf{V}_{h,\mathbf{k}}$, so positive eigenvalues and the corresponding eigenfunctions of Problem 3.1 coincide with ones of Problem 1.2, and so Problem 3.1 is a spectrally correct approximation of Problem 1.1. The only zero eigenfunctions of (3.1) are ones from the kernel of the operator $\nabla_{\mathbf{k}} \times$ i.e. functions from $\nabla_{\mathbf{k}} \, Q_{h,\mathbf{k}}$. If we manage to filter such functions in practice, then the simple formulation (3.1) can be used.

3.1 Eigenvalue solver with projector

The 3D eigenvalue problem results in an algebraic generalized eigenvalue problem with huge matrices. Since we are interested in a few smallest eigenvalues only, an iterative eigenvalue solver is preferable. Such solver usually builds and iteratively improves a search space $\mathrm{span}\{\mathbf{u}_1, \ldots, \mathbf{u}_n\}$, which approximates the smallest eigenvectors. However, there is the drawback, the solver tends to converge first to the eigenvector with the smallest eigenvalue. For the formulation (3.1) it will converge to gradient fields $\nabla_{\mathbf{k}} \, q_h$, $q_h \in Q_{h,\mathbf{k}}$, corresponding to $\lambda_h = 0$. To overcome this behavior one can modify the solver by including a projector to the space, which is orthogonal to discrete gradient field. The approach with a projector is inspired by work [18] and [41].

Assume that there exists a projector $P_{h,\mathbf{k}} \colon \mathbf{X}_{h,\mathbf{k}} \to \mathbf{V}_{h,\mathbf{k}}$, then one can modify the iterative eigenvalue solver in order to construct a search space $\mathrm{span}\{P_{h,\mathbf{k}}\mathbf{u}_1, \ldots, P_{h,\mathbf{k}}\mathbf{u}_N\} \subset \mathbf{V}_{h,\mathbf{k}}$. Since all positive eigenvectors of Problem 3.1 belong to $\mathbf{V}_{h,\mathbf{k}}$, the projector does not affect them, but also does not let the solver converge to eigenfunctions with the eigenvalue zero.

3.2 Projection framework

In Section 3.1 it was mentioned that our formulation of the eigenvalue problem requires an projector $P_{h,\mathbf{k}} \colon \mathbf{X}_h \to \mathbf{V}_{h,\mathbf{k}}$. In order to illustrate the main idea how to construct such a projector, we first describe a projection in the simplified case, for the continuous spaces.

Assume that we have a Helmholtz decomposition in form

$$\mathbf{u} = \nabla_{\mathbf{k}}\, q + \nabla_{\mathbf{k}} \times \mathbf{v},$$

where $\mathbf{u} \in \mathbf{C}^1_{\text{per}}(\Omega)$, $q \in C^2_{\text{per}}(\Omega)$ and $\mathbf{v} \in \mathbf{C}^2_{\text{per}}(\Omega)$. To compute the decomposition we need to find q :

$$-\nabla_{\mathbf{k}} \cdot \mathbf{u} = -\Delta_{\mathbf{k}} q + 0, \tag{3.3}$$
$$q = (-\Delta_{\mathbf{k}})^{-1}(-\nabla_{\mathbf{k}} \cdot \mathbf{u}), \tag{3.4}$$
$$\tilde{\mathbf{u}} = \mathbf{u} - \nabla_{\mathbf{k}}\, q = \nabla_{\mathbf{k}} \times \mathbf{v}. \tag{3.5}$$

Since $b_{\mathbf{k}}(\nabla_{\mathbf{k}} \times \mathbf{v}, f) = 0$ for any $\mathbf{v}$ and $f \in C^2_{\text{per}}(\Omega)$, $\tilde{\mathbf{u}} \in \mathbf{V}_{\mathbf{k}}$ is the projection of $\mathbf{u}$.

Now we are ready to formulate the projector for the modified finite elements. In general, the projection consists of the steps corresponding to (3.3) – (3.5). Operator definitions extensively use terms from Section 2.3.

Lemma 3.1. *The projection $P_{h,\mathbf{k}} \colon \mathbf{X}_{h,\mathbf{k}} \to \mathbf{V}_{h,\mathbf{k}}$ is given by formula*

$$P_{h,\mathbf{k}} = \text{id} - S_{h,\mathbf{k}} \circ C^{-1}_{h,\mathbf{k}} \circ B_{h,\mathbf{k}}$$

and consists of three steps:

1. "div".
   The operator $B_{h,\mathbf{k}} \colon \mathbf{X}_{h,\mathbf{k}} \to Q'_{h,\mathbf{k}}$ is defined as follows, for a given $\mathbf{v}_h \in \mathbf{X}_{h,\mathbf{k}}$ compute $B_{h,\mathbf{k}}\mathbf{v}_h \in Q'_{h,\mathbf{k}}$ by

$$\langle B_{h,\mathbf{k}}\mathbf{v}_h, \phi_{v,\mathbf{k}}\rangle = b_{\mathbf{k}}(\mathbf{v}_h, \phi_{v,\mathbf{k}}), \qquad \text{for all } v \in \mathcal{V}_h. \tag{3.6}$$

2. "Laplace^{-1}".
   The operator $C^{-1}_{h,\mathbf{k}} \colon Q'_{h,\mathbf{k}} \to Q_{h,\mathbf{k}}$ is the inverse of operator $C_{h,\mathbf{k}} \colon Q_{h,\mathbf{k}} \to Q'_{h,\mathbf{k}}$, which is defined as follows, for a given $q_h \in Q_{h,\mathbf{k}}$ compute $C_{h,\mathbf{k}} q_h \in Q'_{h,\mathbf{k}}$ by

$$\langle C_{h,\mathbf{k}} q_h, \phi_{v,\mathbf{k}}\rangle = c_{\mathbf{k}}(q_h, \phi_{v,\mathbf{k}}), \qquad \text{for all } v \in \mathcal{V}_h. \tag{3.7}$$

   The operator $C^{-1}_{h,\mathbf{k}}$ exists, because $c_{\mathbf{k}}(\cdot,\cdot)$ is coercive for $\mathbf{k} \neq \mathbf{0}$ (see Section 1.2).

3. "grad".
   The operator $S_{h,\mathbf{k}} \colon Q_{h,\mathbf{k}} \to \mathbf{X}_{h,\mathbf{k}}$ is defined as follows, for any $q_h \in Q_{h,\mathbf{k}}$ compute $S_{h,\mathbf{k}} q_h = \Pi_{\mathbf{X}_{h,\mathbf{k}}}(\nabla_{\mathbf{k}}\, q_h) \in \mathbf{X}_{h,\mathbf{k}}$, the operator is given by nodal evaluation

$$S_{h,\mathbf{k}} q_h = \sum_{e=(\mathbf{x}_e,\mathbf{y}_e)\in\mathcal{E}_h} \left(\ell_{\mathbf{y}_e,\mathbf{k}}(q_h) e^{i\mathbf{k}\cdot(\mathbf{y}_e - \mathbf{m}_e)} - \ell_{\mathbf{x}_e,\mathbf{k}}(q_h) e^{i\mathbf{k}\cdot(\mathbf{x}_e - \mathbf{m}_e)}\right) \psi_{e,\mathbf{k}}. \tag{3.8}$$

Proof. For any $\mathbf{u}_h \in \mathbf{X}_{h,\mathbf{k}}$, from the discrete Helmholtz decomposition (Lemma 2.6) we have

$$\mathbf{u}_h = \nabla_{\mathbf{k}}\, q_h + \mathbf{v}_h,$$

where $\mathbf{v}_h \in \mathbf{V}_{h,\mathbf{k}}$, $q_h \in Q_{h,\mathbf{k}}$. It follows that for all $f_h \in Q_{h,\mathbf{k}}$

$$(\mathbf{u}_h, \nabla_{\mathbf{k}}\, f_h)_{\mathbf{L}^2} = (\nabla_{\mathbf{k}}\, q_h, \nabla_{\mathbf{k}}\, f_h)_{\mathbf{L}^2} + 0. \tag{3.9}$$

Since (3.9) is true for any $f_h \in Q_{h,\mathbf{k}}$, it is also true for all basis functions $\{\phi_{v,\mathbf{k}}\}$, so (3.9) may be rewritten as following. For all $v \in \mathcal{V}_h$

$$(\nabla_{\mathbf{k}}\, q_h, \nabla_{\mathbf{k}}\, \phi_{v,\mathbf{k}})_{\mathbf{L}^2} = (\mathbf{u}_h, \nabla_{\mathbf{k}}\, \phi_{v,\mathbf{k}})_{\mathbf{L}^2}, \tag{3.10}$$
$$c_{\mathbf{k}}(q_h, \phi_{v,\mathbf{k}}) = b_{\mathbf{k}}(\mathbf{u}_h, \phi_{v,\mathbf{k}}), \tag{3.11}$$
$$\langle C_{h,\mathbf{k}} q_h, \phi_{v,\mathbf{k}}\rangle = \langle B_{h,\mathbf{k}}\mathbf{u}_h, \phi_{v,\mathbf{k}}\rangle. \tag{3.12}$$

Equation (3.12) in operator form gives

$$C_{h,\mathbf{k}} q_h = B_{h,\mathbf{k}} \mathbf{u}_h,$$
$$q_h = C_{h,\mathbf{k}}^{-1} B_{h,\mathbf{k}} \mathbf{u}_h.$$

The operator $C_{h,\mathbf{k}}$ can be inverted because it is positive definite for $\mathbf{k} \neq \mathbf{0}$. Then, having the potential q_h, we can subtract the gradient field $\nabla_{\mathbf{k}} q_h$ from the original $\mathbf{u}_h$, that gives us the orthogonal projection onto $\mathbf{V}_{h,\mathbf{k}}$

$$P_{h,\mathbf{k}} \mathbf{u}_h = (\mathrm{id} - S_{h,\mathbf{k}} \circ C_{h,\mathbf{k}}^{-1} \circ B_{h,\mathbf{k}}) \mathbf{u}_h = (\nabla_{\mathbf{k}} q_h + \mathbf{v}_h) - \nabla_{\mathbf{k}} q_h = \mathbf{v}_h.$$

To prove formula (3.8) we just need to evaluate degrees of freedom using definition (2.3) and applying properties (2.6),

$$S_{h,\mathbf{k}} q_h = \Pi_{\mathbf{X}_{h,\mathbf{k}}} (\nabla_{\mathbf{k}} q_h) = \sum_{e \in \mathcal{E}_h} \ell_{e,\mathbf{k}} (\nabla_{\mathbf{k}} q_h) \psi_{e,\mathbf{k}},$$

$$\begin{aligned}
\ell_{e,\mathbf{k}}(\nabla_{\mathbf{k}} q_h) &= \ell_{e,\mathbf{k}} \left(\nabla_{\mathbf{k}} \sum_{v \in \mathcal{V}_h} \ell_{v,\mathbf{k}}(q_h) \phi_{v,\mathbf{k}} \right) = \ell_{e,\mathbf{k}} \left(\sum_{v \in \mathcal{V}_h} \ell_{v,\mathbf{k}}(q_h) \nabla_{\mathbf{k}} \phi_{v,\mathbf{k}} \right) \\
&= \ell_{e,\mathbf{k}} \left(\sum_{v \in \mathcal{V}_h} \ell_{v,\mathbf{k}}(q_h) \mathrm{e}^{-\mathrm{i}\mathbf{k}\cdot(\mathbf{x}-\mathbf{z}_v)} \nabla \phi_{v,\mathbf{0}} \right) \\
&= \int_{\mathbf{x}_e}^{\mathbf{y}_e} \mathrm{e}^{\mathrm{i}\mathbf{k}\cdot(\mathbf{x}-\mathbf{m}_e)} \left(\sum_{v \in \mathcal{V}_h} \ell_{v,\mathbf{k}}(q_h) \mathrm{e}^{-\mathrm{i}\mathbf{k}\cdot(\mathbf{x}-\mathbf{z}_v)} \nabla \phi_{v,\mathbf{0}} \right) \cdot \mathbf{t}_e ds \\
&= \sum_{v \in \mathcal{V}_h} \ell_{v,\mathbf{k}}(q_h) \mathrm{e}^{\mathrm{i}\mathbf{k}\cdot(\mathbf{z}_v-\mathbf{m}_e)} \int_{\mathbf{x}_e}^{\mathbf{y}_e} \nabla \phi_{v,\mathbf{0}} \cdot \mathbf{t}_e ds \\
&= \sum_{v \in \mathcal{V}_h} \ell_{v,\mathbf{k}}(q_h) \mathrm{e}^{\mathrm{i}\mathbf{k}\cdot(\mathbf{z}_v-\mathbf{m}_e)} (\phi_{v,\mathbf{0}}(\mathbf{y}_e) - \phi_{v,\mathbf{0}}(\mathbf{x}_e)) \\
&= \ell_{\mathbf{y}_e,\mathbf{k}}(q_h) \mathrm{e}^{\mathrm{i}\mathbf{k}\cdot(\mathbf{y}_e-\mathbf{m}_e)} - \ell_{\mathbf{x}_e,\mathbf{k}}(q_h) \mathrm{e}^{\mathrm{i}\mathbf{k}\cdot(\mathbf{x}_e-\mathbf{m}_e)}.
\end{aligned}$$

□

Remark 3.1. Inexact projection.
If $C_{h,\mathbf{k}}^{-1}$ is the exact inversion of the operator $C_{h,\mathbf{k}}$, then the projection is also exact. It follows by construction of the projector.
In practice it may be enough to do an inexact projection. In this case $C_{h,\mathbf{k}}^{-1}$ usually is implemented as an iterative linear solver. For the problem $C_{h,\mathbf{k}} \mathbf{q} = \mathbf{f}$ the residual $\mathbf{r}^n = \mathbf{f} - C_{h,\mathbf{k}} \mathbf{q}^n$ characterizes accuracy of the approximated solution $\mathbf{q}^n$. The stopping criterion is

$$\|\mathbf{r}^n\| < \epsilon_P \|\mathbf{r}^0\|,$$

where $\epsilon_P \in (0, 1)$.

3.2.1 The gradient operator for high order elements

For the high order elements the projection can be done similarly. The only change is that the operator $S_{h,\mathbf{k}}$ is more complex. Here we exploit the local exact sequence property and the fact that by construction $\mathbf{X}_{h,\mathbf{k}}$ explicitly contains gradients of $Q_{h,\mathbf{k}}$.

Let we represent $Q_{h,\mathbf{k}}$ and $\mathbf{X}_{h,\mathbf{k}}$ according to the V-E-F-C (N-E-F-C) basis structure as

$$Q_{h,\mathbf{k}} = Q_V \oplus Q_E \oplus Q_F \oplus Q_C,$$
$$\mathbf{X}_{h,\mathbf{k}} = \mathbf{X}_N \oplus \mathbf{X}_E \oplus \mathbf{X}_{Fg} \oplus \mathbf{X}_{Fn} \oplus \mathbf{X}_{Cg} \oplus \mathbf{X}_{Cn},$$

where Fg and Cg are the parts formed by the gradient basis functions (see Subsection 2.2.4), Fn and Cn are the remaining ones. Assume that the coefficients have the same notations, then the operator $\nabla_{\mathbf{k}}$ can be written in the block form

$$\begin{pmatrix} u_N \\ u_E \\ u_{Fg} \\ u_{Fn} \\ u_{Cg} \\ u_{Cn} \end{pmatrix} = \begin{pmatrix} S_{N,\mathbf{k}} & 0 & 0 & 0 \\ 0 & Id & 0 & 0 \\ 0 & 0 & Id & 0 \\ 0 & 0 & 0 & 0 \\ 0 & 0 & 0 & Id \\ 0 & 0 & 0 & 0 \end{pmatrix} \begin{pmatrix} q_V \\ q_E \\ q_F \\ q_C \end{pmatrix},$$

where (q_V, q_E, q_F, q_C) are the coefficients of a vector from $Q_{h,\mathbf{k}}$, $(u_N, u_E, u_{Fg}, u_{Fn}, u_{Cg}, u_{Cn})$ are the coefficients of the resulting vector in $\mathbf{X}_{h,\mathbf{k}}$, $S_{N,\mathbf{k}}$ is the operator $S_{h,\mathbf{k}}$ for the lowest order elements.

One may see that the operator $S_{h,\mathbf{k}}$ has a very simple sparse structure and so it allows the efficient computation.

3.3 Preconditioned gradient eigenvalue solver

Having a given basis of the finite element space $\mathbf{X}_{h,\mathbf{k}}$ on hand, e.g. $\{\psi_{e,\mathbf{k}}\}$ from Section 2.3, the Problem 3.1 turns into

Problem 3.2. Hermitian generalized matrix eigenvalue problem.
Find a pair of $(\mathbf{u}, \lambda) \in \mathbb{C}^N \times \mathbb{R}^+$ s.t.

$$A\mathbf{u} = \lambda M\mathbf{u},$$

where A and M be $N \times N$ sparse complex matrices corresponding to sesquilinear forms $a_{\mathbf{k}}(\cdot,\cdot)$ and $m(\cdot,\cdot)$, $A = A^H \geq 0$, $M = M^H > 0$.

We are interested in a method approximating the n $(n \ll N)$ smallest eigenvalues and corresponding eigenvectors. For real world 3D Maxwell problems N is quite large, so our eigenvalue problem cannot be solved with a direct method due to huge memory and computational power requirements. This is why we should look for an iterative method for solving such a problem.

Although the matrix A is only positive semidefinite with a huge kernel, for a moment let us assume that A is positive definite. It simplifies the introduction, we will consider the case of a semidefinite matrix later. Assume that there exists a preconditioner T for the operator A for which $\|Id - TA\|_A < 1$ holds. Usually T is applied as an iterative linear solver, and the accuracy of an approximated solution $\mathbf{u}^k$ for the problem $A\mathbf{u} = \mathbf{f}$ is measured by a norm of the residual $\mathbf{r}^k = \mathbf{f} - A\mathbf{u}^k$. The stopping criterion is

$$\|\mathbf{r}^k\| < \epsilon_T \|\mathbf{r}^0\|,$$

where $\epsilon_T \in (0,1)$.

It is well known that the Problem 3.2 has exactly N eigenvalues $0 < \lambda_1 \leq \ldots \leq \lambda_N$ and eigenvectors $\mathbf{u}_1, \ldots, \mathbf{u}_N$, which are M-orthogonal, i.e. $\mathbf{u}_k^H M \mathbf{u}_l = \delta_{kl}$. A good overview of various algebraic eigenvalue problems and methods for solving them can be found in [3].

All stationary points $\mathbf{u}_j$ of the *Rayleigh quotient*

$$\lambda(\mathbf{u}) = \frac{\mathbf{u}^H A\mathbf{u}}{\mathbf{u}^H M\mathbf{u}} \qquad \text{for } \mathbf{u} \neq \mathbf{0}$$

are the eigenvectors $\mathbf{u}_j$ with eigenvalues $\lambda_j = \lambda(\mathbf{u}_j)$, so one may find the eigenvector with the smallest eigenvalue by minimizing the Rayleigh quotient, in particular, a gradient method can be applied.

Since the gradient of the Rayleigh quotient is given by the formula

$$\nabla\lambda(\mathbf{u}) = \frac{2}{\mathbf{u}^H M\mathbf{u}}(A\mathbf{u} - \lambda(\mathbf{u})M\mathbf{u}),$$

we naturally obtain an iterative minimization algorithm in form of a preconditioned gradient method

$$\mathbf{u}^{k+1} = \mathbf{u}^k - \tau^k T(A\mathbf{u}^k - \lambda(\mathbf{u}^k)M\mathbf{u}^k), \tag{3.13}$$

where u^k is a current approximation of the eigenvector, u^{k+1} is the improved one and a parameter $\tau^k > 0$ is selected to improve convergence.

The iterations (3.13) with the choice $\tau^k = 1$ is called *Preconditioned INVerse ITeration* (PINVIT). A version of (3.13) with τ^k minimizing the Rayleigh quotient $\lambda(\mathbf{u}^{k+1})$ is called *preconditioned steepest descent method.* Such preconditioned gradient methods have an important advantage in the implementation, they only require matrix-vector multiplication, a preconditioner routine and almost no extra memory.

[24] proposes the Locally Optimal Block Preconditioned Conjugate Gradient Method based on the three main ideas

- *three-term recurrence for the next approximation*
  Instead of two terms in (3.13) one may use three terms

$$\mathbf{u}^{k+1} = \gamma_1^k \mathbf{w}^k + \gamma_2^k \mathbf{u}^k + \gamma_3^k \mathbf{u}^{k-1},$$
$$\mathbf{w}^k = T(A\mathbf{u}^k - \lambda(\mathbf{u}^k)\, M\mathbf{u}^k),$$

  where $\gamma_1^k, \gamma_2^k, \gamma_3^k$ are some real numbers selected to improve convergence. This scheme is better because it provides more "extrapolation points".

- *use the Rayleigh-Ritz procedure to choose the optimal parameters*
  The goal is to choose $\gamma_1^k, \gamma_2^k, \gamma_3^k$ s.t. they minimize the Rayleigh quotient for $\mathbf{u}^{k+1}$. The problem $A\mathbf{u} = \lambda M\mathbf{u}$ is restricted to the space $\text{span}\{\mathbf{w}^k, \mathbf{u}^k, \mathbf{u}^{k-1}\}$. Let us define 3 by 3 matrices

$$\begin{aligned}\hat{A} &= [\mathbf{w}^k, \mathbf{u}^k, \mathbf{u}^{k-1}]^H A[\mathbf{w}^k, \mathbf{u}^k, \mathbf{u}^{k-1}],\\ \hat{M} &= [\mathbf{w}^k, \mathbf{u}^k, \mathbf{u}^{k-1}]^H M[\mathbf{w}^k, \mathbf{u}^k, \mathbf{u}^{k-1}],\end{aligned}$$

  then we solve a small dense eigenvalue problem $\hat{A}\hat{\mathbf{u}} = \hat{\lambda}\hat{M}\hat{\mathbf{u}}$ with a direct eigenvalue solver. The eigenvector $\hat{\mathbf{u}}_1 = [\gamma_1^k, \gamma_2^k, \gamma_3^k]^T$ corresponding to the smallest eigenvalue $\hat{\lambda}_1$ implicitly provides the optimal choice of the parameters. Setting $\mathbf{u}^{k+1} = [\mathbf{w}^k, \mathbf{u}^k, \mathbf{u}^{k-1}]\hat{\mathbf{u}}_1$, what is actually a sum of $\mathbf{w}^k, \mathbf{u}^k, \mathbf{u}^{k-1}$ with the weights from $\hat{\mathbf{u}}_1$, we provide the minimal Rayleigh quotient available for given $\mathbf{w}^k, \mathbf{u}^k, \mathbf{u}^{k-1}$. The Rayleigh-Ritz method is also used in the preconditioned steepest descent method (3.13), the difference is that only the two-dimensional subspace $[\mathbf{w}^k, \mathbf{u}^k]$ is used and the resulting parameter τ^k can be expressed explicitly by a formula.

- *simultaneous iterations over a block of orthogonal vectors*
  The method we have considered allows to find only the smallest eigenvector. One may keep about the same algorithm, but iterate a block of n M-orthogonal vectors. It requires to change the Rayleigh-Ritz procedure in order to minimize the Rayleigh quotients of all vectors over an extended $3n$-dimensional subspace $[\mathbf{w}_1^k, \mathbf{u}_1^k, \mathbf{u}_1^{k-1}, \ldots, \mathbf{w}_n^k, \mathbf{u}_n^k, \mathbf{u}_n^{k-1}]$. This block algorithm let us find a few first eigenvectors at once. The details will be explained in Section 3.4.

The described three-term recurrence is unstable in machine arithmetic. Step by step $\mathbf{u}^k$ becomes closer and closer to $\mathbf{u}^{k-1}$ that is why the Rayleigh-Ritz method may fail due to the very ill-conditioned matrices $\hat{A}$ and $\hat{M}$. It is better to use the set of vectors $[\mathbf{w}, \mathbf{u}^k, \mathbf{p}^k]$ instead of $[\mathbf{w}, \mathbf{u}^k, \mathbf{u}^{k-1}]$, where $\mathbf{p}^k = \mathbf{u}^k - \gamma_2^{k-1}\mathbf{u}^{k-1}$. The new set is "more orthogonal" and results in the much better conditioned matrices $\hat{A}$ and $\hat{M}$, while $\text{span}\{\mathbf{w}, \mathbf{u}^k, \mathbf{u}^{k-1}\} = \text{span}\{\mathbf{w}, \mathbf{u}^k, \mathbf{p}^k\}$ and so it provides the same three-term recurrence.

An important point is the convergence rate of preconditioned gradient methods, we refer to [24], [31], [32], [25] for an elaborate research and provide just the main result.

Theorem 3.2. *(see [25, Theorem 9])*
Assume that we use the preconditioner $T = T^H > 0$ and let $\kappa(TA)$ be the spectral condition number of the preconditioned operator TA. For a fixed index $j \in \{1, \ldots, n\}$, let λ_j^k be an approximation to the eigenvalue λ_j on step k computed by the block version of the preconditioned steepest descent method. If $\lambda_j^k \in [\lambda_{l_j}, \lambda_{l_j+1})$ then it holds for λ_j^{k+1} that either $\lambda_j^{k+1} < \lambda_{l_j}$ (unless $l_j = j$), or $\lambda_j^{k+1} \in [\lambda_{l_j}, \lambda_j^k)$. In the latter case,

$$\frac{\lambda_j^{k+1} - \lambda_{l_j}}{\lambda_{l_j+1} - \lambda_j^{k+1}} \leq \left(q(\lambda_{l_j}, \lambda_{l_j+1})\right)^2 \frac{\lambda_j^k - \lambda_{l_j}}{\lambda_{l_j+1} - \lambda_j^k},$$

where

$$q(\lambda_{l_j}, \lambda_{l_j+1}) = 1 - \frac{2}{1 + \kappa(TA)}\left(1 - \frac{\lambda_{l_j}}{\lambda_{l_j+1}}\right)$$

is the convergence factor.

The theorem above gives some upper bound for convergence rate of the block preconditioned steepest descent method. At the moment there is no special estimates for the LOBPCG method. Since the LOBPCG method uses the three-term recurrence versus the two-term one for the preconditioned steepest descent method, the Rayleigh-Ritz procedure gives better (smaller) λ_j^k for the LOBPCG. It means that Theorem 3.2 holds true, but the estimate is not sharp enough.

In [24] and [23] there are good numerical experiments concerning practical convergence rate of the LOBPCG. As it was expected, the practical convergence rate is far better and the LOBPCG method may be considered as one of the best among iterative eigenvalue solvers.

3.4 The Projected LOBPCG

To simplify our explanation we have supposed that the matrix A is positive definite, but in fact it is positive definite only on $\mathbf{V}_{h,\mathbf{k}}$ for $\mathbf{k} \neq \mathbf{0}$ and in $\mathbf{X}_{h,\mathbf{k}}$ it has a large kernel. In practice it may be difficult to apply standard preconditioners or direct solvers for such a matrix. We can make the problem a bit easier by using a shift.

Let $A^\delta = A + \delta M$, where $\delta > 0$ is some small regularization parameter. As a result, A^δ is Hermitian and positive definite matrix on the entire $\mathbf{X}_{h,\mathbf{k}}$. Now a lot of preconditioners and direct solvers will work with A^δ. Let T be a preconditioner for the matrix A^δ, in general it may not be Hermitian. All eigenvectors of the problem $A^\delta\mathbf{u} = \lambda^\delta M\mathbf{u}$ are eigenvectors of the original problem $A\mathbf{u} = \lambda M\mathbf{u}$ with the shifted eigenvalues $\lambda = \lambda^\delta - \delta$. So, this shift let us solve the eigenvalue problem with positive definite matrices, while we still can obtain the solution of the original problem.

We modify the LOBPCG algorithm by including the projection $P\colon \mathbf{X}_{h,\mathbf{k}} \to \mathbf{V}_{h,\mathbf{k}}$ described in Section 3.2. The projection is required to overcome the difficulties described in Section 3.1, when an iterative eigenvalue solver tends to converge to eigenvectors for $\lambda = 0$ first. See the Projected LOBPCG algorithm at Figure 3.1, remember that n is the number of eigenvalues we need (a small number).

1. Fill starting vectors $\mathbf{u}_1, \ldots, \mathbf{u}_n$ with random numbers or put some good eigenvector approximations, store them as a matrix $U := [\mathbf{u}_1, \ldots, \mathbf{u}_n]$.
2. Project $\{\mathbf{u}_j\}$ onto $\mathbf{V}_{h,\mathbf{k}}$, $U := PU$.
3. Orthonormalize $\{\mathbf{u}_j\}$ with respect to M-based scalar product using the Gram-Schmidt procedure, s.t. $U^H MU = I$.
4. Construct the restricted matrix, $\hat{A} := U^H AU \quad (\hat{A} \in \mathbb{C}^{n,n})$.
5. Solve dense eigenvalue problem,
$\hat{A}\hat{U} = \hat{U} \operatorname{diag}\{\hat{\lambda}_1, \ldots, \hat{\lambda}_n\} \quad (\hat{U} \in \mathbb{C}^{n,n}, \quad \hat{U}^H \hat{U} = I)$.
6. Construct the first approximations, $U := U\hat{U}$, $\lambda_j := \hat{\lambda}_j$.
7. Calculate residual, $R := AU - MU \operatorname{diag}\{\lambda_1, \ldots, \lambda_n\}$ $(R = [\mathbf{r}_1, \ldots, \mathbf{r}_n])$.
8. If all converged, $\|\mathbf{r}_j\|_{M^{-1}} < \epsilon_E$ for all j, then stop, else continue.
9. Apply preconditioner, $W := TR, \quad T \approx (A + \delta M)^{-1} \quad (W = [\mathbf{w}_1, \ldots, \mathbf{w}_n])$.
10. Project $\{\mathbf{w}_j\}$ onto $\mathbf{V}_{h,\mathbf{k}}$. $W := PW$.
11. Construct the restricted matrices,
$\hat{A} := V^H AV$ and $\hat{M} := V^H MV$ $(\hat{A}, \hat{M} \in \mathbb{C}^{3n,3n}, V = [U, W, Z])$.
12. Solve dense eigenvalue problem, $\hat{A}\hat{V} = \hat{M}\hat{V} \operatorname{diag}\{\hat{\lambda}_1, \ldots, \hat{\lambda}_{3n}\}$
$(\hat{V} = [\hat{U}, \hat{W}, \hat{Z}] \in \mathbb{C}^{3n,3n}$ consists of $\hat{U}, \hat{W}, \hat{Z} \in \mathbb{C}^{3n,n}$, $\hat{V}^H \hat{M} \hat{V} = I)$.
13. Assume $\hat{\lambda}_1 \leq \ldots \leq \hat{\lambda}_{3n}$ and vectors from $\hat{U}$ correspond to $\{\hat{\lambda}_1, \ldots, \hat{\lambda}_n\}$.
Let $\hat{U}_U \in \mathbb{C}^{n,n}$ be the upper part and $\hat{U}_{WZ} \in \mathbb{C}^{2n,n}$ be the lower part of $\hat{U}$, then construct the next approximations,
$Z := [W, Z]\hat{U}_{WZ}$, $U := U\hat{U}_U + Z$, $\lambda_j := \hat{\lambda}_j$, $j = 1, \ldots, n$.
14. Go to 7.

Figure 3.1: Basic algorithm of the LOBPCG eigenvalue solver with the projection.

In practice, we use a more sophisticated algorithm to avoid unnecessary computations and improve stability, the details will be explained later in Section 3.7. Now we comment the main algorithm. The steps 4-6 and 11-13 represent the Rayleigh-Ritz method, in the latter case a wider $3n$-dimensional subspace is used, but only the n smallest eigenvalues and eigenvectors are taken. Moreover, on the very first iteration the matrix Z, which represents $[\mathbf{p}_1, \ldots, \mathbf{p}_n]$ (the method's "memory") is unavailable, so in fact one solves the $2n \times 2n$ subproblem.

Although T is the preconditioner for the matrix A^δ, in the eigensolver we do not use the matrix A^δ itself, but use the matrix A instead. Since the residual can be computed as follows

$$\mathbf{r}_j = (A^\delta - \lambda_j^\delta M)\mathbf{u}_j = (A + \delta M - (\lambda_j + \delta)M)\mathbf{u}_j = (A - \lambda\, M)\mathbf{u}_j,$$

it is more convenient to use A on all steps and not substract δ from resulted eigenvalues. Mathematically it is the same as using A^δ.

In the algorithm we apply the projector only for the starting vectors and the preconditioned residuals W, once per eigensolver iteration. Using mathematical induction we prove that not only W, but all vectors of V belong to $\mathbf{V}_{h,\mathbf{k}}$. Let current U and Z belong to $\mathbf{V}_{h,\mathbf{k}}$, by the definition the new U and Z are just a linear combination of W and the old U and Z, so they also belong to $\mathbf{V}_{h,\mathbf{k}}$.

An important point is when to stop the iteration. How can we control eigenvalue precision? Fortunately it is relatively easy, since $\|\mathbf{u}_j^k\|_M = 1$ (due to the step 5 and 12), then the M^{-1}-norm of the residual $\mathbf{r}_j^k = A\mathbf{u}_j^k - \lambda_j^k M\mathbf{u}_j^k$ and the eigenvalue approximation λ_j^k obtained in the Rayleigh-Ritz procedure give a simple bound for the exact eigenvalue λ_j. According to [34] it is guaranteed that

$$\lambda_j \in [\lambda_j^k - \|\mathbf{r}_j^k\|_{M^{-1}}, \lambda_j^k + \|\mathbf{r}_j^k\|_{M^{-1}}].$$

Moreover, one can use just one Gauss-Seidel step to approximate M^{-1}, because the matrix M is well-conditioned. This yields a quantity which is equivalent to the M^{-1}-norm independent of mesh. Altogether we obtain a simple and efficient termination criterion.

3.5 Discussion of the algorithm

The original LOBPCG algorithm described in [24] and [23] is designed for a positive definite operator. In [41, Section 7.3] it is given a modification of the LOBPCG with the inexact projection, which allows the method to be applied to the Maxwell problem. In Sections 3.4, 3.7 we presented our implementation of the eigensolver. In our opinion, this implementation has some advantages over the one from [41]. We will discuss them in details.

First, we apply the projection only to the vectors W, while [41] applies it to the vectors U and P. It results in doubling of the projection computations.

Second, [41] does not project W, instead it relies on the assumption that the preconditioner $T \approx (A^\delta)^{-1}$ provides $T\mathbf{r} \in \mathbf{V}_{h,\mathbf{k}}$. We know that $(A^\delta)^{-1}\mathbf{r} \in \mathbf{V}_{h,\mathbf{k}}$, but for an inexact inversion this cannot be guaranteed. In [41] the assumption is based on the following points:

- T is a two-level multigrid method resulting from the V-E-F-C splitting of the high order finite elements (the same as we use),
- the coarse level is the lowest order elements and the coarse level correction is solved exactly,
- the fine level is the higher order elements, the smoother is based on the E-F-C splitting with a "reduced" basis (gradient basis functions are ignored).

From these assumptions it does not follow that $T\mathbf{r} \in \mathbf{V}_{h,\mathbf{k}}$, although it could be close and the eigensolver may work. Another weak point is that it requires an exact solving at the coarse

p-multigrid level (the lowest order finite elements). It means that a very fine mesh cannot be treated, because the coarse space is too large. Without the exact solution the method from [41] is unreliable.

3.6 Inexact projection

In Section 3.2 we mentioned that the projection $P\colon \mathbf{X}_{h,\mathbf{k}} \to \mathbf{V}_{h,\mathbf{k}}$ is exact, if the corresponding Poisson problem is solved exactly. Although it is possible, it is not our goal. We are interested in a projection precision as far as it helps to make the eigenvalue solver working. In practice the LOBPCG does not require the exact projection, that we can allow an approximated solution and improve speed.

How precise must the projection be? It is an important question, in order to understand it we shall consider the following example. Assume that we apply an inexact projector $\tilde{P}\colon \mathbf{X}_{h,\mathbf{k}} \to \mathbf{X}_{h,\mathbf{k}}$ s.t. $\tilde{P} \approx P$. It means that for any $\mathbf{w} \in \mathbf{X}_{h,\mathbf{k}}$

$$\tilde{P}\mathbf{w} = \mathbf{u} = \mathbf{v} + \nabla_{\mathbf{k}}\, q,$$

where $\mathbf{v} \in \mathbf{V}_{h,\mathbf{k}}$, $q \in Q_{h,\mathbf{k}}$ and $\|\nabla_{\mathbf{k}}\, q\| \ll 1$. It is the Helmholtz decomposition with a very small gradient field. Now let us try to compute the Rayleigh quotient

$$\tilde{\lambda}(\mathbf{u}) = \frac{\mathbf{u}^H A\mathbf{u}}{\mathbf{u}^H M\mathbf{u}} = \frac{(\varepsilon^{-1}\nabla_{\mathbf{k}}\times\mathbf{u}, \nabla_{\mathbf{k}}\times\mathbf{u})_{\mathbf{L}^2}}{(\mathbf{u},\mathbf{u})_{\mathbf{L}^2}} = \frac{(\varepsilon^{-1}\nabla_{\mathbf{k}}\times\mathbf{v}, \nabla_{\mathbf{k}}\times\mathbf{v})_{\mathbf{L}^2}}{(\mathbf{v},\mathbf{v})_{\mathbf{L}^2} + (\nabla_{\mathbf{k}}q, \nabla_{\mathbf{k}}q)_{\mathbf{L}^2}}.$$

If $\mathbf{v}$ is the exact eigenvector with eigenvalue λ, $\|\mathbf{v}\|_{\mathbf{L}^2} = 1$, then

$$\tilde{\lambda}(\mathbf{u}) = \frac{\lambda}{1 + \|\mathbf{u}-\mathbf{v}\|^2_{\mathbf{L}^2}} = \frac{\lambda}{1 + \|\nabla_{\mathbf{k}}q\|^2_{\mathbf{L}^2}} < \lambda.$$

As we see, in the case of inexact projection the Rayleigh quotient and the Rayleigh-Ritz method always give a bit smaller eigenvalues. It is acceptable since we only look for an approximated solution of the eigenvalue problem and already have some tolerance. But a problem can occur, if the projection error is much greater than this tolerance. The LOBPCG works as follows, it starts with large eigenvalue approximations and then converges down towards the exact eigenvalues until the residual achieves the tolerance. It may happen that due to inexact projection $\tilde{\lambda}$ jumps over λ and continue to go down to zero, to the kernel. To prevent this the projection error must be small enough.

As far as we know, for the LOBPCG with the projection there is no theory to predict safe projection error. In [18] the authors developed such theory, but for the *Projected Preconditioned INverse ITeration* (PPINVIT). Since the Projected LOBPCG may be considered as an extension of the PPINVIT, one can apply the same strategy, but with no guarantee.

The key point of the strategy is to guarantee the following condition on every iteration for some fixed $\epsilon > 0$

$$\frac{\beta}{1-\beta}\frac{\|\mathbf{y} - \tilde{P}\mathbf{y}\|_M}{\|\mathbf{y}\|_A} \le \epsilon, \tag{3.14}$$

where $\mathbf{y}$ is a vector before the projection and $\beta > 0$ is the upper bound of the multigrid convergence rate for the Poisson problem, see [18] for more details. Note that in the algorithm at Figure 3.1 $\mathbf{y} = T\mathbf{r}$. We recommend at least to trace the projection precision according to (3.14), because it is simple and does not require too many computations.

For the LOBPCG algorithm the safe iteration condition is that the Rayleigh-Ritz procedure gives the Rayleigh quotients $\mu_j(\mathbf{u}_j)$ which are not less than the exact eigenvalues λ_j. While this condition is satisfied the iteration continues safely, otherwise some eigenvalues may drop to zero or the algorithm fails due to linear dependency of the iterated vectors. Since the LOBPCG is quite complex, it is hard to prove the safe iteration condition for the full algorithm. But if we assume a non-block version of the algorithm, i.e. just one vector in the iteration, then the safe iteration condition can be formulated as a theorem connecting the eigenvalue solver accuracy and the projection accuracy.

Lemma 3.3 and Theorem 3.4 prove the condition. Although we consider only the first eigenvector, the result can be generalized to other eigenvectors due to the deflation we apply to converged eigenvectors. For the block algorithm one may expect that the theorem gives only the necessary condition and the projection accuracy must be higher.

Lemma 3.3. *Let $\mathbf{u} = \mathbf{v} + \mathbf{u}_g$ be the Helmholtz decomposition, where $\mathbf{v} \in \mathbf{V}_\mathbf{k}$, $\mathbf{u}_g \in \nabla_\mathbf{k} H^1_{\text{per}}(\Omega)$. Assume that $\|\mathbf{u}\| = 1$, $\|\mathbf{u}_g\| \leq \epsilon$, for a small $\epsilon > 0$, also assume that the Rayleigh quotient $\mu(\mathbf{u}) = \langle A_\mathbf{k}\mathbf{u}, \mathbf{u}\rangle$ is s.t. $0 < \mu \leq \lambda_1$, where λ_1 is the minimal positive eigenvalue of $A_\mathbf{k}$. Then the residual can be estimated by $\|A_\mathbf{k}\mathbf{u} - \mu M\mathbf{u}\|_{X'} \leq C_0\epsilon$ with $C_0 = \frac{\lambda_1+\delta}{\sqrt{\delta}}$.*

Proof. We have that

$$\|\mathbf{r}\|^2_{X'} = \|(\mathsf{A}^\delta_\mathbf{k})^{-1}\mathbf{r}\|_X = \|(\mathsf{A}^\delta_\mathbf{k})^{-1}\mathbf{r}\|^2_X = \langle \mathsf{A}^\delta_\mathbf{k}(\mathsf{A}^\delta_\mathbf{k})^{-1}\mathbf{r}, (\mathsf{A}^\delta_\mathbf{k})^{-1}\mathbf{r}\rangle = \langle \mathbf{r}, (\mathsf{A}^\delta_\mathbf{k})^{-1}\mathbf{r}\rangle.$$

Inserting the formula $\mathbf{r} = A_\mathbf{k}\mathbf{u} - \mu M\mathbf{u} = \mathsf{A}^\delta_\mathbf{k}\mathbf{u} - (\mu+\delta)M\mathbf{u}$ we obtain

$$\begin{aligned}
\|\mathbf{r}\|^2_{X'} &= \langle \mathsf{A}^\delta_\mathbf{k}\mathbf{u} - (\mu+\delta)M\mathbf{u}, \mathbf{u} - (\mu+\delta)(\mathsf{A}^\delta_\mathbf{k})^{-1}M\mathbf{u}\rangle \\
&= \langle \mathsf{A}^\delta_\mathbf{k}\mathbf{u}, \mathbf{u}\rangle - 2(\mu+\delta)\langle M\mathbf{u}, \mathbf{u}\rangle + (\mu+\delta)^2\langle \mathbf{u}, (\mathsf{A}^\delta_\mathbf{k})^{-1}M\mathbf{u}\rangle \\
&= (\mu+\delta) - 2(\mu+\delta) + (\mu+\delta)^2\langle \mathbf{u}, (\mathsf{A}^\delta_\mathbf{k})^{-1}M\mathbf{u}\rangle \\
&= (\mu+\delta)\Big[(\mu+\delta)\langle \mathbf{u}, (\mathsf{A}^\delta_\mathbf{k})^{-1}M\mathbf{u}\rangle - 1\Big].
\end{aligned}$$

Now we use the Helmholtz decomposition and let $\mathbf{v} = \sum_j a_j\mathbf{v}_j$ be the representation in eigenfunctions of the operator $A_\mathbf{k}$, then

$$\begin{aligned}
\|\mathbf{r}\|^2_{X'} &= (\mu+\delta)\Big[(\mu+\delta)\left(\frac{1}{\delta}\|\mathbf{u}_g\|^2 + \sum_j \frac{1}{\lambda_j+\delta}a_j^2\right) - 1\Big] \\
&\leq (\lambda_1+\delta)\Big[(\lambda_1+\delta)\left(\frac{1}{\delta}\epsilon^2 + \frac{1}{\lambda_1+\delta}\|\mathbf{u}\|^2\right) - 1\Big] \leq \frac{(\lambda_1+\delta)^2}{\delta}\epsilon^2 = C_0^2\epsilon^2.
\end{aligned}$$

□

Theorem 3.4. *Safe iteration condition.*
Let $\mathbf{u} = \mathbf{v} + \mathbf{u}_g$ be the Helmholtz decomposition, where $\mathbf{v} \in \mathbf{V}_\mathbf{k}$, $\mathbf{u}_g \in \nabla_\mathbf{k} H^1_{\text{per}}(\Omega)$. Assume that $\|\mathbf{u}\| = 1$, $\|\mathbf{u}_g\| \leq \epsilon$ for a small $\epsilon > 0$, also assume that $\|A_\mathbf{k}\mathbf{u} - \mu M\mathbf{u}\|_{X'} > C_0\epsilon$ with $C_0 = \frac{\lambda_1+\delta}{\sqrt{\delta}}$. Then we have $\mu > \lambda_1$.

Proof. If we assume that $\mu \leq \lambda_1$, then Lemma 3.3 gives $\|A_\mathbf{k}\mathbf{u} - \mu M\mathbf{u}\|_{X'} \leq C_0\epsilon$. But this is a contradiction with the assumption $\|A_\mathbf{k}\mathbf{u} - \mu M\mathbf{u}\|_{X'} > C_0\epsilon$, and hence $\mu > \lambda_1$. □

3.7 Implementation

The basic algorithm of the Projected LOBPCG is presented in Figure 3.1. In practice we have many optimizations, some of them improve speed by precomputing and reusing data, others take care of numerical stability. The detailed algorithm is presented in Figure 3.2.

When eigenvectors are converging, the matrices $\hat{A}$ and $\hat{M}$ become more and more ill-conditioned. It happens because $\mathbf{p}^k_j$, $\mathbf{r}^k_j$ and so $\mathbf{w}^k_j$ go to zero, while $\|\mathbf{u}^{k+1}_j - \mathbf{u}^k_j\| \to 0$. Another problem occurs due to the inexact projection, following Section 3.6, if we iterate an already

1. Set the blocksize n and the number of searching vectors g, put $c := 0$, $it := 0$.
2. Fill $U := [\mathbf{u}_1, \dots, \mathbf{u}_n]$ with random vectors or approximations.
3. Project $\{\mathbf{u}_j\}$ onto $\mathbf{V}_{h,\mathbf{k}}$, $U := PU$.
4. Orthonormalize $\{\mathbf{u}_j\}$ by the Gram-Schmidt, s.t. $U^H MU = I$.
5. Construct the restricted matrix, $\hat{A} := U^H AU \quad (\hat{A} \in \mathbb{C}^{n,n})$.
6. Solve dense eigenproblem, $\hat{A}\hat{U} = \hat{U}\,\mathrm{diag}\{\hat{\lambda}_1, \dots, \hat{\lambda}_n\}$ $(\hat{U} \in \mathbb{C}^{n,n}, \hat{U}^H\hat{U} = I)$.
7. Construct the first approximations, $U := U\hat{U}$, $\lambda_j := \hat{\lambda}_j$, $D := \varnothing$.
8. Calculate residual, $R := AU - MU\,\mathrm{diag}\{\lambda_{c+1}, \dots, \lambda_n\}$.
9. If $\{\mathbf{u}_j\}$ $(j = c+1)$ converged, $\|\mathbf{r}_j\|_{M^{-1}} < \epsilon_E$, then $D := [\mathbf{u}_1, \dots, \mathbf{u}_j]$, $c := c+1$, $U := [\mathbf{u}_{c+1}, \dots, \mathbf{u}_n]$, $Z := [\mathbf{z}_{c+1}, \dots, \mathbf{z}_n]$, $R = [\mathbf{r}_{c+1}, \dots, \mathbf{r}_n]$, deflate Z, $Z := Z - D(D^H MZ)$. If $c = g$ exit, else repeat 9 for $j = c+1$.
10. (*optional*) If $c = 1$ select δ s.t. $\delta > -\lambda_1$. If $c > 1$ select $\delta \in (-\lambda_c, -\lambda_{c-1})$.
11. Apply preconditioner, $W := TR, \quad T \approx (A + \delta M)^{-1}$ $(W = [\mathbf{w}_{c+1}, \dots, \mathbf{w}_n])$.
12. Project $\{\mathbf{w}_j\}$ onto $\mathbf{V}_{h,\mathbf{k}}$, $W := PW$, deflate W, $W := W - D(D^H MW)$.
13. If $it = 0$ construct the dense matrices, $\hat{A} := V^H AV$, $\hat{M} := V^H MV$ $(V = [U, W])$, else construct the dense matrices, $\hat{A} := V^H AV$, $\hat{M} := V^H MV$ $(V = [U, W, Z])$.
14. If $it = 0$ solve dense eigenvalue problem, $\hat{A}\hat{V} = \hat{M}\hat{V}\,\mathrm{diag}\{\hat{\lambda}_1, \dots, \hat{\lambda}_{2n}\}$ $(\hat{V} = [\hat{U}, \hat{W}] \in \mathbb{C}^{2n,2n}$ consists of $\hat{U}, \hat{W} \in \mathbb{C}^{2n,n}$, $\hat{V}^H\hat{M}\hat{V} = I)$, else solve dense eigenvalue problem $(l = n - c)$, $\hat{A}\hat{V} = \hat{M}\hat{V}\,\mathrm{diag}\{\hat{\lambda}_{1+3c}, \dots, \hat{\lambda}_{3n}\}$ $(\hat{V} = [\hat{U}, \hat{W}, \hat{Z}] \in \mathbb{C}^{3l,3l}$ consists of $\hat{U}, \hat{W}, \hat{Z} \in \mathbb{C}^{3l,l}$, $\hat{V}^H\hat{M}\hat{V} = I)$.
15. If $it = 0$ set the next approximations, $Z := W\hat{U}_W$, $U := U\hat{U}_U + Z$, $\lambda_j := \hat{\lambda}_j$, $j = c+1, \dots, n$. $\hat{U}_U, \hat{U}_W \in \mathbb{C}^{n,n}$ are the upper and lower part of $\hat{U}$, else set the next approximations, $Z := [W, Z]\hat{U}_{WZ}$, $U := U\hat{U}_U + Z$, $\lambda_j := \hat{\lambda}_j$, $j = c+1, \dots, n$. $\hat{U}_U \in \mathbb{C}^{l,l}$, $\hat{U}_{WZ} \in \mathbb{C}^{2l,l}$ are the upper and lower part of $\hat{U}$.
16. $it := it + 1$. Go to 8.

Figure 3.2: Detailed algorithm of the LOBPCG eigenvalue solver with the projection, deflation and (optional) adaptive δ-shift.

converged eigenvector further, it may "jump over" the exact one and converge to zero. To address these issues one needs to remove $\mathbf{w}_j$ and $\mathbf{p}_j$ corresponding to a converged $\mathbf{u}_j$ from further processing. It should be done without changing the general LOBPCG algorithm.

As a solution we employ the *deflation* workaround to improve both stability and speed. After an eigenvector $\mathbf{u}_j$ has converged ($\|\mathbf{r}(\mathbf{u}_j)\|_{M^{-1}} < \epsilon_E$), it and the corresponding $\mathbf{p}_j$, $\mathbf{w}_j$ are excluded from the iterations and $\mathbf{u}_j$ is moved to a "deflation set" $D = [\mathbf{d}_1, \ldots, \mathbf{d}_c]$, which builds a matrix. So we add one more projection step, every vector from $V = [U, W, Z]$ is replaced with its M-orthogonal projection to $\text{span}\{\mathbf{d}_1, \ldots, \mathbf{d}_c\}^{\perp}$, so it removes all components related to the already converged vectors $\mathbf{u}_1, \ldots, \mathbf{u}_c$ ($\mathbf{d}_1, \ldots, \mathbf{d}_c$), but does not affect the algorithm since all eigenvectors are naturally M-orthogonal. Because of this step the remaining vectors cannot converge to the old ones and the iteration continues safely.

The new projection step is performed by the simple formula $W := W - D(D^H M W)$. It must be applied to the preconditioned residuals $\{\mathbf{w}_j\}$ on every iteration and once to Z, when a new vector has converged. There is no need to deflate U since the vectors are M-orthonormal after the solution of the dense eigenvalue problem, so one can simply remove $\mathbf{u}_j$ from U. Since we iterate progressively less and less vectors the deflation also makes the algorithm faster.

Due to the deflation one should be careful with clustered eigenvalues and the order in which the eigenvectors converge. In the algorithm we require that they converge in increasing order, i.e. even if the residual for λ_{j+1} is below the threshold it is "converged" only after $\lambda_1, \ldots, \lambda_j$ have converged, otherwise the algorithm may fail. Usually this modification gives enough protection.

According to [24], the blocksize n should be more than g, the number of eigenvectors we actually need. In general, the blocksize should be as large as it is needed to contain a possible eigenvalue cluster. One may set $n \approx 2g$, as a start point.

We have stated that the regularization parameter δ must be positive, it guarantees that $A^{\delta} > 0$ and allows us to use any available preconditioner. But some preconditioners can work with an indefinite operator as well. In this case convergence of the eigenvalue solver can be improved. Assume that the lowest positive eigenvalue is λ_1, then the preconditioner $T \approx (A - \tilde{\lambda}_1 M)^{-1}$, where $0 < \tilde{\lambda}_1 < \lambda_1$, may provide faster convergence, i.e. the eigenvalue solver requires less iterations. This can be explained as follows, if $\mathbf{v} = \sum_{j=1}^{N} a_j \mathbf{u}_j$ is a representation in the eigenvector basis of the problem $A\mathbf{u}_j = \lambda_j M \mathbf{u}_j$, then

$$(A - \tilde{\lambda} M)^{-1} \mathbf{v} = \sum_{j=1}^{N} \frac{1}{\lambda_j - \tilde{\lambda}} a_j \mathbf{u}_j.$$

When $\tilde{\lambda}$ is close to some λ_j, then the corresponding eigenvector component is amplified, while the others are diminished. So the preconditioner $T \approx (A - \tilde{\lambda} M)^{-1}$ can be used to improve convergence assuming that all not converged λ_j are greater than $\tilde{\lambda}$ and the preconditioner works well for the indefinite operator. Our multigrid preconditioner (described in Chapter 4) can be applied to the indefinite operator provided that the coarse grid problem has a relatively fine mesh (see [9]).

This observation leads to an "adaptive preconditioner" strategy. We start the computations with $T \approx (A + \delta M)^{-1}$ with some positive δ, after the eigenvalue λ_1 has converged, T is changed to $T \approx (A - \tilde{\lambda}_1 M)^{-1}$, where $\tilde{\lambda}_1$ may be e.g. $0.9\lambda_1$. This step can be used many times and when $\lambda_1, \ldots, \lambda_c$ are converged the preconditioner T may be set to $(A - \tilde{\lambda}_c M)^{-1}$, where $\tilde{\lambda}_c = \lambda_{c-1} + 0.9(\lambda_c - \lambda_{c-1})$. It is hard to predict the exact advantage provided by this step since the effect depends on many factors: quality of the preconditioner, structure of the spectrum, number of the searched eigenvalues and etc. One has to be aware that the effect can be even negative. Optimal parameters should be chosen experimentally and validated in practice.

In case of very ill-conditioned matrices or to get the eigenvectors in high precision it is recommended to apply the Gram-Schmidt orthogonalization to the vectors from V on every iteration. It makes the method more robust, but slower. Since $\hat{M} = I$ the problem $\hat{A}\hat{V} = \hat{M}\hat{V}\,\text{diag}\{\hat{\lambda}_1, \ldots, \hat{\lambda}_{3n}\}$ is simplified to $\hat{A}\hat{V} = \hat{V}\,\text{diag}\{\hat{\lambda}_1, \ldots, \hat{\lambda}_{3n}\}$.

For solving the dense eigenvalue problems in steps 6 and 14 we use the `ZHEGV` routine from the LAPACK software package [1].

Chapter 4

Multigrid as preconditioner

For an efficient solution of the Maxwell eigenvalue problem with the LOBPCG method one first needs an efficient preconditioner for the linear problem.

Problem 4.1. Linear problem.
For some $\delta > 0$ the operator $A^{\delta}_{h,\mathbf{k}} \colon \mathbf{X}_{h,\mathbf{k}} \to \mathbf{X}'_{h,\mathbf{k}}$ is defined by

$$\langle A^{\delta}_{h,\mathbf{k}}\mathbf{u}, \mathbf{v}\rangle = a_{\mathbf{k}}(\mathbf{u}, \mathbf{v}) + \delta\, m(\mathbf{u}, \mathbf{v}), \qquad \mathbf{u}, \mathbf{v} \in \mathbf{X}_{h,\mathbf{k}}.$$

The problem is,
for a given $\mathbf{f} \in \mathbf{X}'_{h,\mathbf{k}}$, find $\tilde{\mathbf{u}} \in \mathbf{X}_{h,\mathbf{k}}$ s.t.

$$\langle A^{\delta}_{h,\mathbf{k}}\tilde{\mathbf{u}}, \mathbf{v}\rangle = \mathbf{f}(\mathbf{v}), \qquad \text{for any } \mathbf{v} \in \mathbf{X}_{h,\mathbf{k}}.$$

A finite element discretization of the linear problem results in a large system with a sparse matrix. Such system usually cannot be solved with a direct linear solver due to huge memory and computational power requirements. Moreover, the LOBPCG method does not require the exact solution of the linear problem, so an iterative method would be the optimal choice. A good overview of various iterative methods for a system of linear equations one can find in [4].

According to [15], [14], [40] multigrid methods provide better convergence rate than other iterative solvers. So, it is desirable to use a multigrid for large scale problems, like our 3D Maxwell problem.

In this chapter we first consider a h-multigrid method for the lowest order finite elements. The last section is devoted to a two-level multigrid method for the higher order elements. Until it is not indicated explicitly we assume $\mathbf{X}_{h,\mathbf{k}}$ and $Q_{h,\mathbf{k}}$ to be the lowest order finite element spaces.

Here we should give a short introduction to multigrid methods, for more detailed information one may refer to [14] and [15]. Let $\tau_0 \subset \ldots \subset \tau_m$ be a sequence of embedded meshes with the corresponding sets of vertices $\mathcal{V}_l$ and edges $\mathcal{E}_l$, where $l \in \{0, \ldots, m\}$. In practice, one usually gets them as a result of successive regular refinement of the original mesh τ_0, then l is the level of refinement and $h_l = O(\frac{h_0}{2^l})$. On the sequence of meshes $\tau_0 \subset \ldots \subset \tau_m$ we can define our finite element spaces $\mathbf{X}_{h,\mathbf{k}}$ and $Q_{h,\mathbf{k}}$, then, by construction, we get a sequence of the embedded spaces $\mathbf{X}_{0,\mathbf{k}} < \ldots < \mathbf{X}_{m,\mathbf{k}}$ and $Q_{0,\mathbf{k}} < \ldots < Q_{m,\mathbf{k}}$. On every finite element space $\mathbf{X}_{l,\mathbf{k}}$ the operator $A^{\delta}_{h,\mathbf{k}}$ is denoted as $A^{\delta}_{l,\mathbf{k}} \colon \mathbf{X}_{l,\mathbf{k}} \to \mathbf{X}'_{l,\mathbf{k}}$. Having an right hand side $\mathbf{f}_l \in \mathbf{X}'_{l,\mathbf{k}}$ we may consider Problem 4.1 on the level l, i.e. find $\tilde{\mathbf{u}}_l$ s.t. $A^{\delta}_{l,\mathbf{k}}\tilde{\mathbf{u}}_l = \mathbf{f}_l$.

In few words, a multigrid is based on the idea that the solution $\tilde{\mathbf{u}}_{l-1}$ on a coarse grid is a pretty good approximation for the solution $\tilde{\mathbf{u}}_l$ on a fine grid, and so information from the level $l-1$ should be used to improve the solution on the level l. Applying the idea recursively one gets a multilevel algorithm, which is very efficient. Any multigrid method consists of the four main parts: interpolation, restriction, smoother and base solver.

To transfer information between coarse and fine meshes we need interpolation and restriction operators. Since we have the embedded sequence of spaces, any $\mathbf{u} \in \mathbf{X}_{l-1,\mathbf{k}}$ is also in $\mathbf{X}_{l,\mathbf{k}}$, to find a copy of $\mathbf{u}$ in $\mathbf{X}_{l,\mathbf{k}}$ we only need to recalculate the degrees of freedom. Recalling the notations for finite elements from Section 2.3, $\psi_1, \ldots, \psi_N$ is a basis of $\mathbf{X}_{l,\mathbf{k}}$ with the corresponding degrees of freedom $\ell_1, \ldots, \ell_N$. The interpolation operator $I_{l-1,l} \colon \mathbf{X}_{l-1,\mathbf{k}} \to \mathbf{X}_{l,\mathbf{k}}$ is defined by formula

$$I_{l-1,l}\mathbf{u} = \sum_{j=1}^{N} \ell_j(\mathbf{u})\psi_j.$$

The restriction operator $R_{l,l-1} \colon \mathbf{X}'_{l,\mathbf{k}} \to \mathbf{X}'_{l-1,\mathbf{k}}$ is defined by $R_{l,l-1} = I^H_{l-1,l}$. Some details concerning the interpolation and restriction operators for the k-modified finite elements are given in Section 4.4.

The next important part of the multigrid method is a smoother, it is an iterative solver with the computational cost $O(N)$ per iteration. A good smoother should efficiently remove the high-frequency part of the error in few iterations. Typical smoothers are stationary iterative methods, e.g. Gauss-Seidel. In Section 4.1 we focus on the problem, which one may have with a smoother for Problem 4.1. In Section 4.3 we address the problem and describe what kind of smoother we use.

The last part of the multigrid is a base solver, it is a solver, which is applied on the coarsest mesh. As the number of unknowns on the coarsest mesh is neglible with respect to the number of unknowns on the finest mesh, one can apply a direct solver to get the exact solution.

4.1 The problem with smoothing

The application of the multigrid method to the Maxwell problem to the $\mathbf{H}(\text{curl})$-conforming finite element discretization is not straightforward. A problem arises because the operator $A^{\delta}_{h,\mathbf{k}}$ of the linear problem scales differently on different components of the Helmholtz decomposition. In that case, standard smoothers like Jacobi or Gauss-Seidel do not provide an appropriate smoothing.

To illustrate this more clearly, let us assume more regularity for a moment and consider the simplified case of Problem 4.1.

Problem 4.2. Associated differential form of Problem 4.1.
Operator $\mathsf{A}^{\delta}_{\mathbf{k}} \colon \mathbf{C}^3_{\text{per}}(\Omega) \to \mathbf{C}^1_{\text{per}}(\Omega)$ is defined by

$$\mathsf{A}^{\delta}_{\mathbf{k}}\mathbf{u} = \nabla_{\mathbf{k}} \times (\varepsilon^{-1} \nabla_{\mathbf{k}} \times \mathbf{u}) + \delta\mathbf{u}.$$

We assume that $\varepsilon \in C^2_{\text{per}}(\Omega)$. The problem is,
for a given $\mathbf{f} \in \mathbf{C}^1_{\text{per}}(\Omega)$ to find $\tilde{\mathbf{u}} \in \mathbf{C}^3_{\text{per}}(\Omega)$ s.t.

$$\mathsf{A}^{\delta}_{\mathbf{k}}\tilde{\mathbf{u}} = \mathbf{f}.$$

Here we mention the differential problem only to better illustrate some simple ideas behind theory and reveal a problem with smoothing.

Let us assume $\varepsilon(\mathbf{x}) = 1$. For $\mathbf{u} \in \mathbf{C}^1_{\text{per}}(\Omega)$ we have the Helmholtz decomposition in form

$$\mathbf{u} = \mathbf{v} + \mathbf{w}, \qquad \mathbf{v} = \nabla_{\mathbf{k}}\, q, \ \mathbf{w} = \nabla_{\mathbf{k}} \times \mathbf{g},$$

where $q \in C^2_{\text{per}}(\Omega)$ and $\mathbf{g} \in \mathbf{C}^2_{\text{per}}(\Omega)$.

Since $\nabla_{\mathbf{k}} \cdot \mathbf{w} = 0$, the differential operator for $\mathbf{w} = \nabla_{\mathbf{k}} \times \mathbf{g}$ gives

$$\mathsf{A}^{\delta}_{\mathbf{k}}\mathbf{w} = \nabla_{\mathbf{k}} \times (\nabla_{\mathbf{k}} \times \mathbf{w}) + \delta\mathbf{w} = -\Delta_{\mathbf{k}}\mathbf{w} + \delta\mathbf{w}.$$

One may see that on a solenoidal field the operator $\mathsf{A}_{\mathbf{k}}^{\delta}$ behaves as a second order elliptic differential operator. The spectrum of $\mathsf{A}_{\mathbf{k}}^{\delta}$ is just the spectrum of the Laplace operator shifted by a small regularization parameter δ.

Since $\nabla_{\mathbf{k}} \times \mathbf{v} = \mathbf{0}$, for the another part of the decomposition $\mathbf{v} = \nabla_{\mathbf{k}} q$ we obtain

$$\mathsf{A}_{\mathbf{k}}^{\delta} \mathbf{v} = \delta \mathbf{v},$$

it is clear that in this case the operator has the absolutely different spectrum. The different behavior of the operator on the parts of the Helmholtz decomposition causes a problem for the multigrid, namely, standard smoothers do not work properly.

As we have seen, the operator $\mathsf{A}_{\mathbf{k}}^{\delta}$ acts as a second order elliptic differential operator on fields of the form $\mathbf{w} = \nabla_{\mathbf{k}} \times \mathbf{g}$. If we manage to restrict to that case, we can reveal multigrid's efficiency. This requires a smoother, which also drops the potential component $\mathbf{v} = -\nabla_{\mathbf{k}} q$, i.e. does projection. Below we first describe an idea how it may be done. For simplicity we work in terms of Problem 4.2.

Let us consider the problem

$$\mathsf{A}_{\mathbf{k}}^{\delta} \mathbf{u} = \nabla_{\mathbf{k}} \times \varepsilon^{-1} \nabla_{\mathbf{k}} \times \mathbf{u} + \delta \mathbf{u} = \mathbf{f}, \qquad \nabla_{\mathbf{k}} \cdot \mathbf{f} = 0.$$

If $\mathbf{u}$ is the exact solution, then one may check that $\nabla_{\mathbf{k}} \cdot \mathbf{u} = 0$ by taking divergence of both sides

$$\nabla_{\mathbf{k}} \cdot (\mathsf{A}_{\mathbf{k}}^{\delta} \mathbf{u}) = \delta \nabla_{\mathbf{k}} \cdot \mathbf{u} = \nabla_{\mathbf{k}} \cdot \mathbf{f} = 0.$$

Now we apply a smoother (e.g. Gauss-Seidel) to Problem 4.2. Since any smoother is just an inexact solver, it only gives the approximated solution $\mathbf{u}$, which may contain both sides of the decomposition

$$\mathbf{u} = \mathbf{v} + \mathbf{w}, \qquad \mathbf{v} = \nabla_{\mathbf{k}} q,\ \mathbf{w} = \nabla_{\mathbf{k}} \times \mathbf{g}.$$

Now we would like to remove the potential component $\mathbf{v}$, to do that it is enough to calculate q. Taking into account that $\nabla_{\mathbf{k}} \times \mathbf{v} = \mathbf{0}$ and $\nabla_{\mathbf{k}} \cdot \mathbf{w} = 0$ we obtain

$$\mathbf{r} = \mathbf{f} - \mathsf{A}_{\mathbf{k}}^{\delta} \mathbf{u} = \mathbf{f} - \nabla_{\mathbf{k}} \times \varepsilon^{-1} \nabla_{\mathbf{k}} \times \mathbf{w} - \delta(\mathbf{v} + \mathbf{w}), \tag{4.1}$$

$$-\nabla_{\mathbf{k}} \cdot \mathbf{r} = \delta \nabla_{\mathbf{k}} \cdot \mathbf{v} = \delta \nabla_{\mathbf{k}} \cdot \nabla_{\mathbf{k}} q = \delta \Delta_{\mathbf{k}} q, \tag{4.2}$$

$$-q = (-\delta \Delta_{\mathbf{k}})^{-1}(-\nabla_{\mathbf{k}} \cdot \mathbf{r}), \tag{4.3}$$

$$\mathbf{w} = \mathbf{u} - \mathbf{v} = \mathbf{u} - \nabla_{\mathbf{k}} q = \mathbf{u} + \nabla_{\mathbf{k}}(-\delta \Delta_{\mathbf{k}})^{-1}(-\nabla_{\mathbf{k}} \cdot \mathbf{r}). \tag{4.4}$$

As we see, to remove the potential component completely it requires the exact solution of the Poisson problem. Although, in general it is possible, the computational cost is too high in context of smoothing.

A compromise could be applying one step of a stationary iterative solver (e.g. Gauss-Seidel) to find an approximated solution of the Poisson problem. This approach gives us a composite smoother.

4.2 Multilevel nodal decomposition

Now we go back to Problem 4.1. To construct a correct smoother we follow an approach presented in [16], an alternative approach is described in [2]. Here we need the Helmholtz decomposition, a property that we do have for $\mathbf{X}_{l,\mathbf{k}}$.

The idea is the following: we represent $\mathbf{X}_{m,\mathbf{k}}$, our finite element space on the finest level as a multilevel nodal decomposition

$$\mathbf{X}_{m,\mathbf{k}} = \mathbf{X}_{0,\mathbf{k}} + \sum_{l=1}^{m} \sum_{e \in \mathcal{E}_l} \operatorname{span}\{\psi_{e,\mathbf{k}}\} + \sum_{l=1}^{m} \sum_{v \in \mathcal{V}_l} \operatorname{span}\{\nabla_{\mathbf{k}} \phi_{v,\mathbf{k}}\}. \tag{4.5}$$

The part with $\text{span}\{\nabla_{\mathbf{k}}\,\phi_{v,\mathbf{k}}\}$ represents the potential part of the Helmholtz decomposition, while the part with $\text{span}\{\psi_{e,\mathbf{k}}\}$ represents the solenoidal part. It is a hint that the *multiplicative Schwarz framework* of (4.5) gives us a multigrid V-cycle with a sort of Gauss-Seidel smoother. Some fundamentals of a multilevel nodal decomposition and its connection with multigrid and Gauss-Seidel is well explained in [40].

First we have to prove that this decomposition guarantees a sufficient decoupling of subspaces in terms of energy, independent of m. To show this we need to prove two properties.

Let us formally denote the subspaces from (4.5) as $\mathbf{Y}_j^l$, then we can rewrite (4.5) in form

$$\mathbf{X}_{m,\mathbf{k}} = \mathbf{Y}^0 + \sum_{l=1}^{m}\sum_{e\in\mathcal{E}_l}\mathbf{Y}_e^l + \sum_{l=1}^{m}\sum_{v\in\mathcal{V}_l}\mathbf{Y}_v^l,$$

where $\{\mathbf{Y}_n\}$ denotes the set of all $\mathbf{Y}_e^l$ and $\mathbf{Y}_v^l$, so that any $\mathbf{Y}_n$ corresponds to some $\mathbf{Y}_e^l$, $\mathbf{Y}_v^l$ or $\mathbf{Y}^0$.

The first property to be proven is a *stability estimate*

$$\inf\left\{\sum_n \|\mathbf{v}_n\|^2_{A^\delta_{m,\mathbf{k}}} \mid \sum_n \mathbf{v}_n = \mathbf{v}, \mathbf{v}_n \in \mathbf{Y}_n\right\} \le C_{\text{stab}}\|\mathbf{v}\|^2_{A^\delta_{m,\mathbf{k}}} \quad \text{for all } \mathbf{v}\in\mathbf{X}_{m,\mathbf{k}}. \tag{4.6}$$

We need to define subspaces $\{\mathbf{U}_n\}$. First we distribute all basis functions $\{\psi_{e,\mathbf{k}}\}$ of $\mathbf{X}_{0,\mathbf{k}},\dots,\mathbf{X}_{m,\mathbf{k}}$ and $\{\nabla_{\mathbf{k}}\,\phi_{v,\mathbf{k}}\}$ of $\nabla_{\mathbf{k}}\,Q_{0,\mathbf{k}},\dots,\nabla_{\mathbf{k}}\,Q_{m,\mathbf{k}}$ among small number of classes, s.t. the supports of the functions inside every particular class are mutually nonoverlapping. Then, building the span of all functions for each class we get the subspaces $\{\mathbf{U}_n\}$. On each level of refinement l a fixed small number of such $\mathbf{U}_n$ is enough to cover the whole space $\mathbf{X}_{l,\mathbf{k}}$.

The second property to be proven is a *strengthened Cauchy-Schwarz inequality*

$$|\langle A^\delta_{m,\mathbf{k}}\mathbf{v}_j, \mathbf{v}_n\rangle| \le C_{\text{orth}}\gamma^{|j-n|}\|\mathbf{v}_j\|_{A^\delta_{m,\mathbf{k}}}\|\mathbf{v}_n\|_{A^\delta_{m,\mathbf{k}}}, \tag{4.7}$$

for all $\mathbf{v}_j\in\mathbf{U}_j, \mathbf{v}_n\in\mathbf{U}_n$ and some γ s.t. $0\le\gamma<1$.

It is important that C_{stab} and C_{orth} have to be independent of h and m.

Theorem 4.1. *(see [16, Theorem 3.1])*
Provided that properties (4.6) *and* (4.7) *hold, the convergence rate ρ of the multigrid V-cycle in norm $\|\cdot\|_{A^\delta_{m,\mathbf{k}}}$ is bounded by*

$$\rho \le 1 - \frac{1}{C_{\text{stab}}(1+\sigma)^2} \quad \textit{with} \quad \sigma = C_{\text{orth}}\frac{1+\gamma}{1-\gamma}.$$

In [16] it was proven that the properties (4.6) and (4.7), and hence Theorem 4.1 are valid for the standard Nédélec finite elements ($\mathbf{k}=\mathbf{0}$) and Dirichlet boundary condition. The arguments can also be transfered to the modified finite elements.

4.3 The hybrid smoother

Let $S_{h,\mathbf{k}}\colon Q_{h,\mathbf{k}}\to\mathbf{X}_{h,\mathbf{k}}$ and $C_{h,\mathbf{k}}\colon Q_{h,\mathbf{k}}\to Q'_{h,\mathbf{k}}$ be the gradient and Laplacian operators defined in Section 3.2, $A^\delta_{h,\mathbf{k}}\colon\mathbf{X}_{h,\mathbf{k}}\to\mathbf{X}'_{h,\mathbf{k}}$ be the problem operator. Following [40] we derive a Gauss-Seidel type smoother as a *successive subspace correction* algorithm for the space decomposition

$$\mathbf{X}_{h,\mathbf{k}} = \sum_{e\in\mathcal{E}}\text{span}\{\psi_{e,\mathbf{k}}\} + \sum_{v\in\mathcal{V}}\text{span}\{\nabla_{\mathbf{k}}\,\phi_{v,\mathbf{k}}\} = \sum_{j=1}^{|\mathcal{E}|}\mathbf{Y}_j + \sum_{j=|\mathcal{E}|+1}^{|\mathcal{E}|+|\mathcal{V}|}\mathbf{Y}_j. \tag{4.8}$$

Note that (4.8) is just one layer of the multilevel nodal decomposition (4.5) corresponding to some level l. Therefore, the SSC algorithm for (4.8), which builds a smoother, is a part of the

1. Calculate the first residual, $\mathbf{r}^1 = \mathbf{f} - A^\delta_{h,\mathbf{k}}\mathbf{x}^0$.

2. Do steps 3–5 for $n = 1, \ldots, N$.

3. Solve the correction equation in the subspace $\mathbf{Y}_n$:
find $\mathbf{c}^n \in \mathbf{Y}_n$ s.t. $\langle A^\delta_{h,\mathbf{k}}\mathbf{c}^n, \mathbf{v}\rangle = \mathbf{r}^n(\mathbf{v})$ for all $\mathbf{v} \in \mathbf{Y}_n$.

4. Update the approximation, $\mathbf{x}^n = \mathbf{x}^{n-1} + \mathbf{c}^n$.

5. Update the residual, $\mathbf{r}^{n+1} = \mathbf{f} - A^\delta_{h,\mathbf{k}}\mathbf{x}^n = \mathbf{r}^n - A^\delta_{h,\mathbf{k}}\mathbf{c}^n$.

Figure 4.1: The SSC algorithm for the linear problem $A^\delta_{h,\mathbf{k}}\mathbf{x} = \mathbf{f}$ and a space decomposition $\mathbf{X}_{h,\mathbf{k}} = \sum_n \mathbf{Y}_n$.

SSC algorithm for (4.5), which builds multigrid with the smoother. The SSC iterations are described in Figure 4.1.

Denote $a_{ij} = \langle A^\delta_{h,\mathbf{k}}\psi_{e_i,\mathbf{k}}, \psi_{e_j,\mathbf{k}}\rangle$ to be the matrix entries of the operator $A^\delta_{h,\mathbf{k}}$ in the basis $\{\psi_{e,\mathbf{k}}\}$ of $\mathbf{X}_{h,\mathbf{k}}$, $f_j = \mathbf{f}(\psi_{e_j,\mathbf{k}})$ and x_j be the coordinate vectors of $\mathbf{f}$ and $\mathbf{x}$ in the same basis. The step 3 of the SSC algorithm for the subspace $\mathbf{Y}_n = \operatorname{span}\{\psi_{e_n,\mathbf{k}}\}$ gives

$$x_n a_{nn} = \langle A^\delta_{h,\mathbf{k}}(x_n\psi_{e_n,\mathbf{k}}), \psi_{e_n,\mathbf{k}}\rangle = \langle \mathbf{r}^n, \psi_{e_n,\mathbf{k}}\rangle = \langle \mathbf{f} - A^\delta_{h,\mathbf{k}}\mathbf{x}^{n-1}, \psi_{e_n,\mathbf{k}}\rangle$$

or

$$x_n = a_{nn}^{-1}(f_n - \sum_{j=1}^{n-1} a_{nj}x_j), \qquad n = 1, \ldots, N,$$

so it is just Gauss-Seidel for the matrix a_{ij}.

Next we proceed with the SSC algorithm for the second part of (4.8). From the previous part we have the current approximation $\mathbf{x} \in \mathbf{X}_{h,\mathbf{k}}$ and the residual $\mathbf{r} \in \mathbf{X}'_{h,\mathbf{k}}$, and so we continue to solve the correction equations. On the second part of (4.8), in the space $\nabla_\mathbf{k} Q_{h,\mathbf{k}}$ we may see that

$$\langle A^\delta_{h,\mathbf{k}} \nabla_\mathbf{k} \phi_{v_i,\mathbf{k}}, \nabla_\mathbf{k} \phi_{v_j,\mathbf{k}}\rangle = 0 + \delta\, m(\nabla_\mathbf{k} \phi_{v_i,\mathbf{k}}, \nabla_\mathbf{k} \phi_{v_j,\mathbf{k}}) = \delta\langle C_{h,\mathbf{k}}\phi_{v_i,\mathbf{k}}, \phi_{v_j,\mathbf{k}}\rangle,$$
$$\mathbf{r}(\nabla_\mathbf{k} \phi_{v_j,\mathbf{k}}) = \langle \mathbf{r}, \nabla_\mathbf{k} \phi_{v_j,\mathbf{k}}\rangle = \langle S^H_{h,\mathbf{k}}\mathbf{r}, \phi_{v_j,\mathbf{k}}\rangle,$$

therefore, the linear problem $A^\delta_{h,\mathbf{k}}\mathbf{u} = \mathbf{r}$ transforms into $\delta C_{h,\mathbf{k}} q = S^H_{h,\mathbf{k}}\mathbf{r}$, where $\mathbf{u} = \nabla_\mathbf{k}\, q$.

Denote $c_{ij} = \delta\langle C_{h,\mathbf{k}}\phi_{v_i,\mathbf{k}}, \phi_{v_j,\mathbf{k}}\rangle$ to be the matrix entries of the operator $\delta C_{h,\mathbf{k}}$ in the basis $\{\phi_{v,\mathbf{k}}\}$ of $Q_{h,\mathbf{k}}$, $d_j = \langle S^H_{h,\mathbf{k}}\mathbf{r}, \phi_{v_j,\mathbf{k}}\rangle$ and q_j be the coordinate vectors of $S^H_{h,\mathbf{k}}\mathbf{r}$ and $\mathbf{q}$ in the same basis. Then the SSC algorithm in the subspace $\nabla_\mathbf{k} Q_{h,\mathbf{k}}$ gives

$$q_n = c_{nn}^{-1}(d_n - \sum_{j=1}^{n-1} c_{nj}q_j), \qquad n = 1, \ldots, N,$$

and we again obtain Gauss-Seidel, but for the matrix (c_{ij}). Taking into account the last correction, the complete result of the SSC algorithm for the whole decomposition (4.8) is given by $\tilde{\mathbf{x}} = \mathbf{x} + S_{h,\mathbf{k}}q$.

Let $\mathbf{c} = GS(A)\mathbf{r}$ denotes the Gauss-Seidel preconditioner for a linear system $A\mathbf{c} = \mathbf{r}$, namely, let U_A be the upper triangular part of the matrix A, then $GS(A)\mathbf{r} = U_A^{-1}\mathbf{r}$. Now we define a hybrid smoother $\mathbf{c} = HS^\nu_h(\mathbf{r})$, where ν is a number of iterations. See the algorithm in Figure 4.2. The steps 3–5 of the algorithm correspond to (4.2)–(4.4).

The smoother works in the following way. As an input it takes the current residual $\mathbf{r} \in \mathbf{X}'_{h,\mathbf{k}}$, $\mathbf{r} = \mathbf{f} - A^\delta_{h,\mathbf{k}}\mathbf{u}^0$, where $\mathbf{u}^0$ is the current approximated solution. As a result it returns a correction

1. Initialization, $\mathbf{c} = \mathbf{0}$, $\mathbf{f} = \mathbf{r}$, $it = 1$.
2. Standard smoother, $\mathbf{d} = GS(A^{\delta}_{h,\mathbf{k}})\mathbf{f}$.
3. Restrict to the potential space, $q = S^{H}_{h,\mathbf{k}}(\mathbf{f} - A^{\delta}_{h,\mathbf{k}}\mathbf{d})$.
4. Potential smoother, $p = GS(\delta\, C_{h,\mathbf{k}})q$.
5. Return from the potential space, $\mathbf{d} = \mathbf{d} + S_{h,\mathbf{k}}\, p$.
6. Apply the correction, $\mathbf{c} = \mathbf{c} + \mathbf{d}$.
7. Update the residual, $\mathbf{f} = \mathbf{f} - A^{\delta}_{h,\mathbf{k}}\mathbf{d}$.
8. While $it < \nu$ do $it = it + 1$ and go to 2.

Figure 4.2: Algorithm of the hybrid smoother $\mathbf{c} = HS^{\nu}(\mathbf{r})$.

$\mathbf{c} \in \mathbf{X}_{h,\mathbf{k}}$, s.t. $\mathbf{u}^0 + \mathbf{c}$ is an improved approximated solution. In an operator form the hybrid smoother is

$$\begin{aligned} HS^1_h =& GS(A^{\delta}_{h,\mathbf{k}}) + S_{h,\mathbf{k}} \circ GS(\delta\, C_{h,\mathbf{k}}) \circ S^{H}_{h,\mathbf{k}} \circ (Id - A^{\delta}_{h,\mathbf{k}} \circ GS(A^{\delta}_{h,\mathbf{k}})) \\ =& GS(A^{\delta}_{h,\mathbf{k}}) + S_{h,\mathbf{k}} \circ GS(\delta\, C_{h,\mathbf{k}}) \circ S^{H}_{h,\mathbf{k}} \\ &- S_{h,\mathbf{k}} \circ GS(\delta\, C_{h,\mathbf{k}}) \circ S^{H}_{h,\mathbf{k}} \circ A^{\delta}_{h,\mathbf{k}} \circ GS(A^{\delta}_{h,\mathbf{k}}). \end{aligned} \tag{4.9}$$

It should be mentioned that the hybrid smoother is not symmetric, even if one uses Symmetric Gauss-Seidel instead of *GS*. It is clear from (4.9), if *GS* were symmetric, then the first two items are symmetric, but the last one is definitely not. This fact brakes symmetry of whole multigrid framework.

There is a way to restore symmetry. We need to construct the adjoint operator for HS^{ν}_l, let us name it AHS^{ν}_l. As we will see later in Section 4.5 we need both HS^{ν}_l and AHS^{ν}_l to make symmetric multigrid. In few words, to get AHS^{ν}_l one has to apply the SSC algorithm in inverse order. In practice it means to swap order of the standard and potential smoothers and use the adjoint smoothers. See the algorithm in Figure 4.3. Here $\mathbf{c} = GS^H(A)\mathbf{r}$ denotes the adjoint version of Gauss-Seidel, namely, $GS^H(A) = (U^H_A)^{-1}$, where U_A is the upper triangular part of the matrix A.

4.4 Interpolation and restriction operators

Since we use nonstandard degrees of freedom, to be consistent we need to modify intermesh operators as well. Let $\mathbf{X}_{l,\mathbf{k}}$ and $Q_{l,\mathbf{k}}$ be the finite element spaces $\mathbf{X}_{h,\mathbf{k}}$ and $Q_{h,\mathbf{k}}$ on a h-refinement level l. Usually one constructs the interpolation operator $I_{l-1,l}\colon \mathbf{X}_{l-1,\mathbf{k}} \to \mathbf{X}_{l,\mathbf{k}}$, then the restriction operator $R_{l,l-1}\colon \mathbf{X}'_{l,\mathbf{k}} \to \mathbf{X}'_{l-1,\mathbf{k}}$ is obtained by $R_{l,l-1} = I^H_{l-1,l}$.

For our k-modified finite elements, the operators $I_{l-1,l}$ and $R_{l,l-1}$ are also k-dependent. Since $\mathbf{X}_{l-1,\mathbf{k}}$ is a subspace of $\mathbf{X}_{l,\mathbf{k}}$, to set up the interpolation operator it is enough to calculate fine degrees of freedom for coarse nodal bases. Those values form interpolation weights, which are coefficients of the operator $I_{l-1,l}$.

Here one should recall the notations for finite elements from Section 2.3. For simplicity, let us consider an edge $e \in \mathcal{E}_l$ and its refined edges $e_1, e_2 \in \mathcal{E}_{l+1}$ of the equal length. They have the associated degrees of freedom $\ell_{e,\mathbf{k}}, \ell_{e_1,\mathbf{k}}, \ell_{e_2,\mathbf{k}}$, the nodal points $\mathbf{m}_e, \mathbf{m}_{e_1}, \mathbf{m}_{e_2}$ (midpoints of the edges) and the nodal bases $\psi_{e,\mathbf{k}}, \psi_{e_1,\mathbf{k}}, \psi_{e_2,\mathbf{k}}$. Using the definition of degree of freedom we

1. Initialization, $\mathbf{c} = \mathbf{0}$, $\mathbf{f} = \mathbf{r}$, $it = 1$.
2. Restrict to the potential space, $q = S_{h,\mathbf{k}}^H \mathbf{f}$.
3. Potential smoother, $p = GS^H(\delta C_{h,\mathbf{k}})q$.
4. Return from the potential space, $\mathbf{d} = S_{h,\mathbf{k}}\, p$.
5. Standard smoother, $\mathbf{d} = \mathbf{d} + GS^H(A_{h,\mathbf{k}}^\delta)(\mathbf{f} - A_{h,\mathbf{k}}^\delta \mathbf{d})$.
6. Apply the correction, $\mathbf{c} = \mathbf{c} + \mathbf{d}$.
7. Update the residual, $\mathbf{f} = \mathbf{f} - A_{h,\mathbf{k}}^\delta \mathbf{d}$.
8. While $it < \nu$ do $it = it + 1$ and go to 2.

Figure 4.3: Algorithm of the adjoint hybrid smoother $\mathbf{c} = AHS^\nu(\mathbf{r})$.

obtain, for $\mathbf{k} = \mathbf{0}$

$$\ell_{e1,\mathbf{0}}(\psi_{e,\mathbf{0}}) = \ell_{e2,\mathbf{0}}(\psi_{e,\mathbf{0}}) = 0.5\ell_{e,\mathbf{0}}(\psi_{e,\mathbf{0}}) = 0.5,$$

for $\mathbf{k} \neq \mathbf{0}$

$$\begin{aligned}
\ell_{e1,\mathbf{k}}(\psi_{e,\mathbf{k}}) = \ell_{e1,\mathbf{k}}(\mathrm{e}^{-\mathrm{i}\mathbf{k}\cdot(\mathbf{x}-\mathbf{m}_e)}\psi_{e,\mathbf{0}}) &= \int_{e1} \mathrm{e}^{\mathrm{i}\mathbf{k}\cdot(\mathbf{x}-\mathbf{m}_{e_1})}\mathrm{e}^{-\mathrm{i}\mathbf{k}\cdot(\mathbf{x}-\mathbf{m}_e)}\psi_{e,\mathbf{0}} \cdot \mathbf{t}_{e_1} ds \\
&= \mathrm{e}^{\mathrm{i}\mathbf{k}\cdot(\mathbf{m}_e-\mathbf{m}_{e_1})} \int_{e1} \psi_{e,\mathbf{0}} \cdot \mathbf{t}_{e_1} ds = \mathrm{e}^{\mathrm{i}\mathbf{k}\cdot(\mathbf{m}_e-\mathbf{m}_{e_1})}\ell_{e1,\mathbf{0}}(\psi_{e,\mathbf{0}}) \\
&= 0.5\mathrm{e}^{\mathrm{i}\mathbf{k}\cdot(\mathbf{m}_e-\mathbf{m}_{e_1})},
\end{aligned}$$

by analogy

$$\ell_{e_2,\mathbf{k}}(\psi_{e,\mathbf{k}}) = 0.5\mathrm{e}^{\mathrm{i}\mathbf{k}\cdot(\mathbf{m}_e-\mathbf{m}_{e_2})}.$$

Similar changes are required for the interpolation and restriction operators in $Q_{h,\mathbf{k}}$ space. For simplicity we consider an edge $e \in \mathcal{E}_l$ with two vertices $v_1, v_2 \in \mathcal{V}_l$. After refinement we get vertices $v_{11}, v_{22}, v_{33} \in \mathcal{V}_{l+1}$ on the fine mesh, their coordinates are

$$\mathbf{z}_{v_{11}} = \mathbf{z}_{v_1},\ \mathbf{z}_{v_{22}} = \mathbf{z}_{v_2},\ \mathbf{z}_{v_{12}} = 0.5(\mathbf{z}_{v_1} + \mathbf{z}_{v_2}) = \mathbf{m}_e.$$

All vertices have the corresponding degrees of freedom

$$\ell_{v_1,\mathbf{k}}, \ell_{v_2,\mathbf{k}}, \ell_{v_{11},\mathbf{k}}, \ell_{v_{22},\mathbf{k}}, \ell_{v_{12},\mathbf{k}}$$

and the nodal bases

$$\phi_{v_1,\mathbf{k}}, \phi_{v_2,\mathbf{k}}, \phi_{v_{11},\mathbf{k}}, \phi_{v_{22},\mathbf{k}}, \phi_{v_{12},\mathbf{k}}.$$

Simple calculations give, for $\mathbf{k} = \mathbf{0}$

$$\ell_{v_{12},\mathbf{0}}(\phi_{v_1,\mathbf{0}}) = \ell_{v_{12},\mathbf{0}}(\phi_{v_2,\mathbf{0}}) = 0.5\ell_{v_1,\mathbf{0}}(\phi_{v_1,\mathbf{0}}) = 0.5\ell_{v_2,\mathbf{0}}(\phi_{v_2,\mathbf{0}}) = 0.5,$$

for $\mathbf{k} \neq \mathbf{0}$

$$\begin{aligned}
\ell_{v_{12},\mathbf{k}}(\phi_{v_1,\mathbf{k}}) &= \ell_{v_{12},\mathbf{k}}(\mathrm{e}^{-\mathrm{i}\mathbf{k}\cdot(\mathbf{x}-\mathbf{z}_{v_1})}\phi_{v_1,\mathbf{0}}) = \mathrm{e}^{-\mathrm{i}\mathbf{k}\cdot(\mathbf{z}_{v_{12}}-\mathbf{z}_{v_1})}\phi_{v_1,\mathbf{0}}(\mathbf{z}_{v_{12}}) \\
&= \mathrm{e}^{-\mathrm{i}\mathbf{k}\cdot(\mathbf{z}_{v_{12}}-\mathbf{z}_{v_1})}\ell_{v_{12},\mathbf{0}}(\phi_{v_1,\mathbf{0}}) = 0.5\mathrm{e}^{\mathrm{i}\mathbf{k}\cdot(\mathbf{z}_{v_1}-\mathbf{z}_{v_{12}})}, \\
\ell_{v_{11},\mathbf{k}}(\phi_{v_1,\mathbf{k}}) &= \phi_{v_1,\mathbf{k}}(\mathbf{z}_{v_{11}}) = \phi_{v_1,\mathbf{k}}(\mathbf{z}_{v_1}) = 1,
\end{aligned}$$

by analogy

$$\ell_{v12,\mathbf{k}}(\phi_{v2,\mathbf{k}}) = 0.5e^{i\mathbf{k}\cdot(\mathbf{z}_{v2}-\mathbf{z}_{v12})},$$
$$\ell_{v22,\mathbf{k}}(\phi_{v2,\mathbf{k}}) = 1.$$

Similar formulas can be derived for other cases of interpolation. One may notice that the interpolation weights include a phase shift factor related to shift (if any) between the nodal points of coarse and fine finite element spaces.

4.5 The h-multigrid implementation

We have described all parts needed for construction a multigrid framework. Multigrid may be implemented as a linear solver, but we prefer to use it as a preconditioner together with Krylov-subspace iterative solvers, e.g. GMRes, BiCGStab or CG.

By $\mathbf{c}_l = HS_l^{\nu}(\mathbf{r}_l)$ we denote ν iterations of the hybrid smoother on the refinement level l, the smoother is defined in Section 4.3. The interpolation and restriction operators $I_{l-1,l}$ and $R_{l,l-1}$ were also defined there. From these parts we construct a recursive *coarse grid correction $V(\nu_1, \nu_2)$-cycle* beginning on a level l with the right hand side $\mathbf{r}_l$. We denote it as $\mathbf{c}_l = MG_l^{\nu_1,\nu_2}(\mathbf{r}_l)$, it takes the current residual $\mathbf{r}_l \in \mathbf{X}'_{l,\mathbf{k}}$, $\mathbf{r}_l = \mathbf{f}_l - A^{\delta}_{l,\mathbf{k}}\mathbf{u}_l^0$, where $\mathbf{u}_l^0$ is the current approximated solution. As a result it returns a correction $\mathbf{c}_l \in \mathbf{X}_{l,\mathbf{k}}$ s.t. $\mathbf{u}_l^0 + \mathbf{c}_l$ is an improved approximated solution. See the algorithm in Figure 4.4.

Since the operators $A^{\delta}_{l,\mathbf{k}}$ and $C_{l,\mathbf{k}}$ are Hermitian we may exploit this fact to construct a symmetric multigrid preconditioner. Such preconditioner can be combined with iterative solvers designed for symmetric operators, e.g. CG method. If a pre-smoother on the step 1 of the multigrid algorithm is the adjoint operator of the post-smoother on the step 7, then the multigrid V(ν, ν)-cycle is symmetric. For instance, one may use HS_l^{ν} as a pre-smoother and AHS_l^{ν} as a post-smoother.

We also apply the multigrid algorithm for solving the Poison problem in the projector from Section 3.2. Again, the multigrid is used as a preconditioner for Krylov-subspace iterative solvers. The difference is that everything happens in spaces $Q_{l,\mathbf{k}}$, where there is no problem with smoothing, and so no need to use non-standard smoother, like the hybrid one. In that case a few iterations of the standard Gauss-Seidel $\mathbf{c} = GS(\mathbf{r})$ smoother or its symmetric version are used, moreover, the operator $(\delta\, C_{l,\mathbf{k}})$ is used instead of $A^{\delta}_{l,\mathbf{k}}$.

4.6 Two-level multigrid for high order finite elements

Let us consider Problem 4.1 on higher order finite elements. Since we use the hierarchical finite elements there is a sequence of embedded spaces, higher order spaces include lower order spaces. So the successive subspace correction method can be applied to this sequence also. By construction the high order finite element space $\mathbf{X}_{h,\mathbf{k}}$ has the natural N-E-F-C decomposition

$$\mathbf{X}_{h,\mathbf{k}} = \mathbf{X}_N \oplus \mathbf{X}_E \oplus \mathbf{X}_F \oplus \mathbf{X}_C,$$

which can be used as a basis for a multilevel method.

In practice such a multilevel method is often realized as a two-level method, where $\mathbf{X}_{h,\mathbf{k}}$ is the fine space and $\mathbf{X}_N$ is the coarse space. Although it is possible to use more than two levels it is not efficient. The lowest order space $\mathbf{X}_N$ plays a special role, it carries the global information. To solve the problem on the coarse level we can either use the h-multigrid described above, or apply a direct solver when the problem size is not so large. Since the basis of $\mathbf{X}_{h,\mathbf{k}}$ explicitly includes the basis of $\mathbf{X}_N$ the interlevel interpolation $I : \mathbf{X}_N \to \mathbf{X}_{h,\mathbf{k}}$ just means to ignore the higher order degrees of freedom and identify the rest ones. The restriction $R : \mathbf{X}'_{h,\mathbf{k}} \to \mathbf{X}'_N$ is

1. Pre-smoothing, $\mathbf{c}_l^1 = HS_l^{\nu_1}(\mathbf{r}_l)$.
2. Update the residual, $\mathbf{f}_l = \mathbf{r}_l - A_{l,\mathbf{k}}^{\delta}\mathbf{c}_l^1$.
3. Restrict the residual, $\mathbf{f}_{l-1} = R_{l,l-1}\mathbf{f}_l$.
4. Coarse grid correction,
   If $l > 1$ then $\mathbf{c}_{l-1} = MG_{l-1}^{\nu_1,\nu_2}(\mathbf{f}_{l-1})$,
   else solve $A_{0,\mathbf{k}}^{\delta}\mathbf{c}_0 = \mathbf{f}_0$ with a direct solver.
5. Interpolate the correction, $\mathbf{c}_l^2 = I_{l-1,l}\mathbf{c}_{l-1}$.
6. Update the residual, $\mathbf{f}_l = \mathbf{f}_l - A_{l,\mathbf{k}}^{\delta}\mathbf{c}_l^2$.
7. Post-smoothing, $\mathbf{c}_l^3 = HS_l^{\nu_2}(\mathbf{f}_l)$.
8. Sum the corrections, $\mathbf{u}_l = \mathbf{c}_l^1 + \mathbf{c}_l^2 + \mathbf{c}_l^3$.

Figure 4.4: Algorithm of multigrid preconditioner $\mathbf{c}_l = MG_l^{\nu_1,\nu_2}(\mathbf{r}_l)$.

the adjoint of I. The smoother can be implemented in different ways depending on the exact form of the decomposition.

There are some important results available in the theory of *additive Schwarz methods* (ASM) which prove properties of such a smoother. We mention them without going into details, we refer to [38] and [35] for a systematic explanation.

Lemma 4.2. *Parameter-robust ASM preconditioner for* $\mathbf{H}(\text{curl})$.
(we refer to [41, Corollary 6.5]).
Let $\mathbf{X}_{h,p} \subset \mathbf{H}(\text{curl}, \Omega)$ *denote a Nédélec finite element space of order* p *and* $Q_{h,p+1} \subset \mathbf{H}^1(\Omega)$ *an appropriate scalar finite element space of order* $p+1$ *with* $\ker(\text{curl}, \mathbf{X}_{h,p}) = \nabla Q_{h,p+1}$ *and* $\nabla Q_{h,p+1} \subset \mathbf{X}_{h,p}$.
We consider the following subspace splitting of the finite element spaces, assuming finite overlap (N_0 denotes maximal number of the overlapping spaces):

$$Q_{h,p+1} = \textstyle\sum_i^{M_Q} Q_i \quad \textit{and} \quad \mathbf{X}_{h,p} = \sum_j^{M_X} \mathbf{X}_j.$$

If for any Q_i *there exists* $\mathbf{X}_j$ *s.t.* $\nabla Q_i \subset \mathbf{X}_j$ *and if the local splitting provides the estimates*

$$\inf_{\{\mathbf{u}:\, \mathbf{u}=\sum \mathbf{u}_j, \mathbf{u}_j \in \mathbf{X}_j\}} \left\{ \sum_j^{M_X} \|\mathbf{u}_j\|_{\mathbf{H}(\text{curl})}^2 \right\} \le c_1(h,p)\|\mathbf{u}\|_{\mathbf{H}(\text{curl})}^2,$$

$$\inf_{\{q:\, q=\sum q_i, q_i \in Q_i\}} \left\{ \sum_i^{M_Q} \|\nabla q_i\|_0^2 \right\} \le c_2(h,p)\|\nabla q\|_0^2,$$

then the additive Schwarz preconditioner C *built on the space splitting* $\{\mathbf{X}_j\}$*, applied to the parameter-dependent Problem 4.1, is robust with respect to the parameter* δ*. For any* $\mathbf{u} \in \mathbf{X}_{h,p}$

$$(c_1(h,p) + c_2(h,p)(1+c_F)^2)^{-1}\langle \mathsf{C}\mathbf{u}, \mathbf{u}\rangle \le \langle A_{h,\mathbf{k}}^{\delta}\mathbf{u}, \mathbf{u}\rangle \le N_0\langle \mathsf{C}\mathbf{u}, \mathbf{u}\rangle.$$

Note that here the preconditioner C *is s.t.* C^{-1} *approximates* $(A_{h,\mathbf{k}}^{\delta})^{-1}$.

This lemma applied to our particular construction of the finite element spaces gives the corollary.

Corollary 4.1. (see [41, Corollary 6.6])

1. Let the finite element spaces $\mathbf{X}_{h,p} \subset \mathbf{H}(\mathrm{curl},\Omega)$ and $Q_{h,p+1} \subset H^1(\Omega)$ satisfy the local exact sequence property (Lemma 2.4). Then any space splitting $\{\mathbf{X}_j\}$ of the space $\mathbf{X}_{h,p}$ built on the (possibly finer) N-E-F-C splitting

$$\mathbf{X}_{h,p} = \mathbf{X}_{h,N} \oplus \sum_{e\in\mathcal{E}_h} \mathbf{X}_{E_e} \oplus \sum_{f\in\mathcal{F}_h} \mathbf{X}_{F_f} \oplus \sum_{c\in\mathcal{C}_h} \mathbf{X}_{C_c}$$

specifies a parameter-robust ASM preconditioner.

2. Let $\mathbf{X}_{h,p} \subset \mathbf{H}(\mathrm{curl},\Omega)$ and $Q_{h,p+1} \subset H^1(\Omega)$ be defined as it was done in Subsection 2.2.3 and 2.2.4, i.e. the basis of $\mathbf{X}_{h,p}$ explicitly contains the gradients of the higher order basis functions of $Q_{h,p}$. Then any space splitting $\mathbf{X}_{h,p} = \sum_j \mathbf{X}_j$ which is built on a two-level concept, where the correction on the lowest order space $\mathbf{X}_{h,N}$ is either solved exactly or is done by h-multigrid based on the *Arnold-Falk-Winther* splitting or on the one of Hiptmair (4.8) implies a parameter-robust ASM preconditioner.

Let us consider an overlapping decomposition of $\mathbf{X}_{h,\mathbf{k}}$

$$\mathbf{X}_{h,\mathbf{k}} = \mathbf{X}_N \oplus \mathbf{X}_E \oplus \mathbf{X}_F \oplus \mathbf{X}_C. \tag{4.10}$$

For the given decomposition the SSC algorithm results in a series of Gauss-Seidel smoothers applied respectively to the edge-, face- and cell-based basis functions. Finally on the high level we have a smoother consisting of 3 Gauss-Seidel subsmoothers related to the E-F-C parts, which are applied consecutively in 3 subspace correction iterations. On the lowest level (space $\mathbf{X}_N$) we either use a direct solver or apply already described h-multigrid with the hybrid smoother. Since the hybrid smoother respects the Hiptmair splitting, the two-level framework gives us a parameter-robust ASM preconditioner.

The two-level multigrid concept above does not require any special changes in case of the modified elements. The interpolation and restriction operators between the lowest and higher order spaces stay as they are due to their simplicity. For the high order smoother we just need to assemble k-modified matrices, that was already explained. The implementation of h-multigrid was also described above.

Chapter 5

Scalable parallelization model

Multigrid methods provide the most efficient way to solve large linear systems arising from the finite element discretization of the Maxwell equations. But our final goal, the band structure computation for a given 3D material distribution, is much more challenging since we have to solve multiple eigenvalue problems, each of them requires the preconditioning and the projection on every step. Total computational requirements are too high for a personal computer, but could be meet by modern parallel supercomputers. A software implementation must employ parallel algorithms and appropriate data structures in order to run on a parallel computer. Moreover, in our case the software implementation also has to provide a realization of the periodic boundary condition. Here we present an approach for constructing an efficient parallel multigrid implementation.

Our parallelization model follows the concept presented in [39]. The model is realized in a free open source parallel finite element software package M++ written in *programming language C++* with an extensive use of the *generic programming* paradigm. The key point of the concept is geometric-centered data structures, which allow to avoid any *global numbering* issues by identifying parallel distributed objects by their geometric position. The parallelization is done for the *distributed memory model* assuming an implementation via the Message Passing Interface (MPI) software library. The geometric-centered data structures in M++ provide a very good *parallel scalability* up to 1000 processors and more.

In this chapter we briefly describe main points essential for the parallelization and a realization of the periodic finite elements.

5.1 Parallel mesh model

For the construction we employ the concept of *Distributed Point Objects*, which states that any element of a mesh, like a vertex, edge, face and cell, is associated with a geometric point in $\mathbb{R}^3$ which acts as a global identification key. All objects of the mesh are stored in *hash map containers* with position as a key. Although a global geometric ordering may be introduced and tree-based map containers could also be used, a hash map provides faster access requiring only $O(1)$ operations. Since floating point computations on a computer assume some rounding error one has to introduce some geometric tolerance ϵ_g and take it into account in comparing the positions.

Let $\mathcal{C}_l, \mathcal{V}_l, \mathcal{E}_l, \mathcal{F}_l$ define correspondingly sets of cells, vertices, edges and faces on a level $l = 0, \dots, m$. On the coarsest level $l = 0$ one defines a problem statement with a geometry, material distribution and the boundary conditions. Successive regular refinements give us level $l = m$, where the problem to be solved, as well as a number of intermediate levels used for the multigrid method. The computational domain $\overline{\Omega}_l$ is decomposed as follows $\overline{\Omega}_l = \bigcup_{c \in \mathcal{C}_l} \overline{\Omega}_c$, where Ω_c is the interior of the cell.

Let $c \in \mathcal{C}_l$ be a cell on level l, then define $\mathcal{V}_c \subset \mathbb{R}^3, \mathcal{E}_c, \mathcal{F}_c$ to be the sets of vertices, edges and faces, which belong to the cell. Every edge $e \in \mathcal{E}_c$ is defined by a pair of vertices $(\mathbf{x}_e, \mathbf{y}_e)$ and is associated with the midpoint $\mathbf{z}_e = 0.5(\mathbf{x}_e + \mathbf{y}_e)$. By analogy all faces and cells are defined by an ordered set of vertices and are associated with the corresponding midpoints. An interior face $f \in \mathcal{F}_l$ is also associated with the pair of cell midpoints $(\mathbf{z}_{c_1}, \mathbf{z}_{c_2})$ s.t. $f \in \mathcal{F}_{c_1} \cap \mathcal{F}_{c_2}$, boundary faces are associated with the pair $(\mathbf{z}_c, \infty)$, where ∞ is a special point. The vertices and midpoints build the set of cell hash keys $\mathcal{Z}_c = \mathcal{V}_c \cup \{\mathbf{z}_e \colon e \in \mathcal{E}_c\} \cup \{\mathbf{z}_f \colon f \in \mathcal{F}_c\} \cup \{\mathbf{z}_c\}$.

For the parallelization it is needed to distribute the mesh among processes. A *load balancing* procedure should do it in such a way to provide about equal amount of work for all processes. We will not focus on this point, but assume that the parallel distribution is done on the coarsest level and is inherited on finer levels. This may be described as follows, let $\mathcal{P} = \{1, \ldots, P\}$ define the process set, then the distribution is given by a function $\mathrm{dest} \colon \mathcal{C}_0 \to \mathcal{P}$. The function specifies a base non-overlapping decomposition of cells

$$\mathcal{C}_0 = \bigcup_{p \in \mathcal{P}} \mathcal{C}_0^p, \qquad \text{where } \mathcal{C}_0^p = \{c \in \mathcal{C}_0 \colon \mathrm{dest}(c) = p\}.$$

It induces an overlapping decomposition of the domains $\overline{\Omega}^p = \bigcup_{c \in \mathcal{C}_0^p} \overline{\Omega}_c$, the vertices $\mathcal{V}_0^p = \bigcup_{c \in \mathcal{C}_0^p} \mathcal{V}_c$, edges $\mathcal{E}_0^p = \bigcup_{c \in \mathcal{C}_0^p} \mathcal{E}_c$, faces $\mathcal{F}_0^p = \bigcup_{c \in \mathcal{C}_0^p} \mathcal{F}_c$ and so the hash keys $\mathcal{Z}_0^p = \bigcup_{c \in \mathcal{C}_0^p} \mathcal{Z}_c$. The overlapping decomposition of the hash keys $\mathcal{Z}_0$ defines a set-valued *partition map*

$$\pi_0 : \mathcal{Z}_0 \to 2^P \quad \text{s.t.} \quad \pi_0(\mathbf{z}) = \{p \in \mathcal{P} \colon \mathbf{z} \in \mathcal{Z}_0^p\}.$$

The parallel decomposition of all data is done by the main process and then the results $\mathcal{C}_0^p, \mathcal{V}_0^p, \mathcal{E}_0^p, \mathcal{F}_0^p, \mathcal{Z}_0^p$ and $\pi_0^p = \pi_0|_{\mathcal{Z}_0^p}$ are distributed among other processes.

The next step, successive regular refinements can be performed in parallel by all processes and does not require intercommunication. During the refinement from level l to level $l+1$ on a process p, a cell $c \in \mathcal{C}_l^p$ is divided into few cells C_c, forming a new set of cells $\mathcal{C}_{l+1}^p = \bigcup_{c \in \mathcal{C}_l^p} C_c$. It also induces the new sets $\mathcal{V}_{l+1}^p, \mathcal{E}_{l+1}^p, \mathcal{F}_{l+1}^p, \mathcal{Z}_{l+1}^p$. New vertices, edges, faces and cells on level $l+1$ inherit the partition map from the elements of level l they located on, e.g. a new vertex $\mathbf{x} \in \mathcal{V}_{l+1}^p$ located on the edge $e \in \mathcal{E}_l^p$ has $\pi_{l+1}^p(\mathbf{x}) = \pi_l^p(\mathbf{z}_e)$. An important point is that the local refinement must be consistent across the parallel boundaries

$$\pi_l^p(\mathbf{z}) = \pi_l^q(\mathbf{z}), \quad \text{for any } \mathbf{z} \in \mathcal{Z}_l^p \cap \mathcal{Z}_l^q \text{ and } l = 0, \ldots, m,$$

the regular refinement holds this property.

5.2 Parallel finite elements

The finite elements are build over the mesh objects, so they inherit the parallel domain decomposition. Let $\mathbf{X}_l = \mathrm{span}\{\phi_{l,j} \colon j \in \mathcal{I}_l\}$ be a global finite element space on level l, where $\{\phi_{l,j}\}$ is a basis and $\mathcal{I}_l$ is the index of basis functions. Note that the global basis is not available directly, but it could be assembled of shape functions. For every basis function $\phi_{l,j} \in \mathbf{X}_l$ there exists the dual function $\phi'_{l,j} \in \mathbf{X}'_l$ s.t. $\langle \phi'_{l,i}, \phi_{l,j} \rangle = \delta_{i,j}$, the dual functions also are not available directly.

Each index $j \in \mathcal{I}_l$ is associated with a unique nodal point $\mathbf{n}_j \in \overline{\Omega}_l$, which is located on the mesh element corresponding to the basis function $\phi_{l,j}$, e.g. a nodal point is located on an edge, if its function is edge-based. Let $\mathcal{N}_l = \{\mathbf{n}_j \colon j \in \mathcal{I}_l\}$ be the set of nodal points. In practice every process stores only a hash set of local nodal points $\mathcal{N}_l^p = \{\mathbf{n}_j \in \mathcal{N}_l \colon \mathbf{n}_j \in \overline{\Omega}^p\}$, the set is constructed in parallel by all processes, going along all cells $\mathcal{C}_l^p$ and mapping the nodal points from the reference element. Since every element of the mesh already has the partition map

$\pi_l(\mathbf{z})$ we can define the partition map for a nodal point $\pi_l(\mathbf{n}_j) = \pi_l(j) = \pi_l(\mathbf{z})$, where $\mathbf{z}$ is the midpoint of the mesh element. It induces an overlapping decomposition of the indices

$$\mathcal{I}_l = \bigcup_{p\in\mathcal{P}} \mathcal{I}_l^p, \qquad \text{where } \mathcal{I}_l^p = \{j \in \mathcal{I}_l \colon p \in \pi_l(j)\}.$$

Set N_l^p to be the number of indices in $\mathcal{I}_l^p$. There is no need for a fixed global or local numbering inside the indices, any numbering may be used, because the unique identification is provided by the geometric position $\mathbf{n}_j$.

A finite element function $\mathbf{v}_l = \sum_{j\in\mathcal{I}_l} v_{l,j}\phi_{l,j} \in \mathbf{X}_l$ is represented uniquely by its coefficient vector $\tilde{\mathbf{v}}_l = (v_{l,j})_{j\in\mathcal{I}_l}$, with $v_{l,j} = \langle\phi'_{l,j}, \mathbf{v}_l\rangle$. By analogy a discrete functional $\mathbf{f}_l = \sum_{j\in\mathcal{I}_l} f_{l,j}\phi'_{l,j} \in \mathbf{X}'_l$ is represented uniquely by the coefficient vector $\tilde{\mathbf{f}}_l = (f_{l,j})_{j\in\mathcal{I}_l}$, with $f_{l,j} = \langle\mathbf{f}_l, \phi_{l,j}\rangle$. The spaces $\mathbf{X}_l$ and $\mathbf{X}'_l$ should be treated separately, so we need different representations for them.

Let us define an auxiliary space of distributed vectors

$$\underline{\mathbf{X}}_l^{\mathcal{P}} = \{(\mathbf{u}_l^1, \dots, \mathbf{u}_l^P) \mid \mathbf{u}_l^p \in \mathbb{C}^{N_l^p}, p \in \mathcal{P}\}.$$

The coefficient vector $\tilde{\mathbf{v}}_l$ is represented in parallel by its local restrictions $\underline{\mathbf{v}}_l^p = (v_{l,j})_{j\in\mathcal{I}_l^p} \in \mathbb{C}^{N_l^p}$ stored on the process p. It defines a mapping from a vector (identified with the coefficient vector) to the distributed vector

$$E_l \colon \mathbf{X}_l \to \underline{\mathbf{X}}_l^{\mathcal{P}} \qquad \text{s.t. } \mathbf{v}_l \to \underline{\mathbf{v}}_l = (\underline{\mathbf{v}}_l^1, \dots, \underline{\mathbf{v}}_l^P).$$

By definition this mapping is unique and consistent, i.e. the coefficients coincide on the parallel interfaces. We define a space of consistent distributed vectors

$$\underline{\mathbf{X}}_l = \{\underline{\mathbf{v}}_l \in \underline{\mathbf{X}}_l^{\mathcal{P}} \mid v_{l,j}^p = v_{l,j}^q \text{ for any } p,q,j \text{ s.t. } p,q \in \pi_l(j)\}.$$

So, any vector $\mathbf{v}_l \in \mathbf{X}_l$ has the unique and consistent parallel representation $\underline{\mathbf{v}}_l = E_l(\mathbf{v}_l) \in \underline{\mathbf{X}}_l$.

Right-hand side vectors and matrices are also distributed, in parallel each process assembles only its local contribution corresponding to a parallel domain $\overline{\Omega^p}$. For $\mathbf{f}_l \in \mathbf{X}'_l$ every process computes only the local contribution $\underline{\mathbf{f}}_l^p = (f_{l,j})_{j\in\mathcal{I}_l^p} \in \mathbb{C}^{N_l^p}$. The full coefficient vector $\tilde{\mathbf{f}}_l$ can be obtained after a parallel collection and has the additive decomposition

$$f_{l,j} = \sum_{p\in\pi_l(j)} f_{l,j}^p. \tag{5.1}$$

It means that for functionals in $\mathbf{X}'_l$ we use a non-unique representation in the quotient space $\underline{\mathbf{X}}'_l = \underline{\mathbf{X}}_l^{\mathcal{P}}/\underline{\mathbf{X}}_l^0$, where the collection kernel $\underline{\mathbf{X}}_l^0$ of the operation (5.1) is given by

$$\underline{\mathbf{X}}_l^0 = \Big\{\underline{\mathbf{f}}_l \in \underline{\mathbf{X}}_l^{\mathcal{P}} \mid \sum_{p\in\pi_l(j)} f_{l,j}^p = 0, \quad j \in \mathcal{I}_l\Big\}.$$

The parallel inner product is defined on $\underline{\mathbf{X}}_l^{\mathcal{P}} \times \underline{\mathbf{X}}_l^{\mathcal{P}}$

$$(\underline{\mathbf{f}}_l, \underline{\mathbf{v}}_l) = \sum_{p\in\mathcal{P}} (\underline{\mathbf{f}}_l^p, \underline{\mathbf{v}}_l^p) = \sum_{j\in\mathcal{I}_l} v_{l,j}\overline{f_{l,j}}, \qquad \underline{\mathbf{u}}_l, \underline{\mathbf{v}}_l \in \underline{\mathbf{X}}_l^{\mathcal{P}}.$$

The inner product can also be considered as a dual pairing on $\underline{\mathbf{X}}'_l \times \underline{\mathbf{X}}_l$, then the kernel space $\underline{\mathbf{X}}_l^0$ may be represented as a *polar space*

$$\underline{\mathbf{X}}_l^0 = \{\underline{\mathbf{v}}_l \in \underline{\mathbf{X}}_l^{\mathcal{P}} \mid (\underline{\mathbf{v}}_l, \underline{\mathbf{w}}_l) = 0, \quad \underline{\mathbf{w}}_l \in \underline{\mathbf{X}}_l\}.$$

With respect to the dual pairing we obtain the adjoint operator for E_l, $E'_l \colon \underline{\mathbf{X}}'_l \to \mathbf{X}'_l$

$$E'_l(\underline{\mathbf{f}}_l) = \mathbf{f}_l = \sum_{p\in\mathcal{P}} \sum_{j\in\mathcal{I}_l^p} f_{l,j}^p \phi'_{l,j}.$$

Note that this mapping is not one-to-one. An unique parallel representation of functionals is provided by a non-overlapping decomposition of the index set $\mathcal{I}_l = \bigcup_{p\in\mathcal{P}} \hat{\mathcal{I}}_l^p$, where

$$\hat{\mathcal{I}}_l^p = \{j \in \mathcal{I}_l^p \mid p = \min \pi_l(j)\}.$$

If an index j belongs to few processes, then the process with the minimal number is the master process, only it stores data for the index. As a result we get a subspace of $\underline{\mathbf{X}}_l'$

$$\hat{\underline{\mathbf{X}}}_l' = \{\underline{\mathbf{f}}_l \in \underline{\mathbf{X}}_l' \mid f_{l,j}^p = 0 \quad \text{for all } j \notin \hat{\mathcal{I}}_l^p, \quad j \in \mathcal{I}_l\},$$

Now E_l' restricted to $\hat{\underline{\mathbf{X}}}_l'$ is one-to-one.

For different numerical aspects we need to define norms. Since the coefficient vectors are distributed definition of norms is not a simple task and may require special efforts.

A simple norm in the space $\underline{\mathbf{X}}_l^{\mathcal{P}}$ may be defined as follows

$$\|\underline{\mathbf{v}}_l\|_{\underline{\mathbf{X}}^{\mathcal{P}}} = \sqrt{\sum_{p\in\mathcal{P}} \sum_{j\in\mathcal{I}_l^p} |v_{l,j}^p|^2} = \sqrt{\sum_{p\in\mathcal{P}} \|\underline{\mathbf{v}}_l^p\|^2}, \qquad \underline{\mathbf{v}}_l \in \underline{\mathbf{X}}_l^{\mathcal{P}}.$$

The norm has no physical meaning, it is mesh-dependent and not invariant with respect to the parallel distribution. But evaluation of the norm is very straightforward, just compute the standard norm of the local parts in parallel and take a sum. That is why this norm can be used in cases when the exact value is not important, e.g. for contolling the convergence of an iterative method.

For a consistent vector in $\underline{\mathbf{X}}_l$ we define the norm

$$\|\underline{\mathbf{v}}_l\|_{\underline{\mathbf{X}}_l} = \sqrt{\sum_{p\in\mathcal{P}} \sum_{j\in\hat{\mathcal{I}}_l^p} |v_{l,j}^p|^2}, \qquad \underline{\mathbf{v}}_l \in \underline{\mathbf{X}}_l,$$

which is invariant of the load balancing. For a functional in $\underline{\mathbf{X}}_l'$ we define the dual norm

$$\|\underline{\mathbf{f}}_l\|_{\underline{\mathbf{X}}_l'} = \sup_{\|\underline{\mathbf{v}}_l\|_{\underline{\mathbf{X}}_l}=1} (\underline{\mathbf{f}}_l, \underline{\mathbf{v}}_l), \qquad \underline{\mathbf{f}}_l \in \underline{\mathbf{X}}_l'.$$

The norm requires a parallel communication in order to compute the unique representation $\hat{\underline{\mathbf{f}}}_l \in \hat{\underline{\mathbf{X}}}_l'$ of $\underline{\mathbf{f}}_l$ and then

$$\|\underline{\mathbf{f}}_l\|_{\underline{\mathbf{X}}_l'} = \|\hat{\underline{\mathbf{f}}}_l\|_{\underline{\mathbf{X}}_l'} = \|\hat{\underline{\mathbf{f}}}_l\|_{\underline{\mathbf{X}}_l^{\mathcal{P}}} = \sqrt{\sum_{p\in\mathcal{P}} \|\hat{\underline{\mathbf{f}}}_l^p\|^2}.$$

5.3 Operator representation

Let $a(\cdot,\cdot)\colon \mathbf{X}_l \times \mathbf{X}_l \to \mathbb{C}$ be an sesquilinear form, which induces a discrete operator $A_l\colon \mathbf{X}_l \to \mathbf{X}_l'$. We assume that the sesquilinear form $a(\cdot,\cdot)$ allows for a cell-based additive decomposition

$$a(\mathbf{u}, \mathbf{v}) = \sum_{c\in\mathcal{C}_l} a_c(\mathbf{u}, \mathbf{v}), \qquad \mathbf{u}, \mathbf{v} \in \mathbf{X}_l,$$

and assume that the global basis functions $\{\phi_{l,j}\}$ of the finite elements have local support s.t. $a_c(\phi_{l,j}, \phi_{l,m}) \neq 0$ only if their nodal points $\mathbf{n}_j, \mathbf{n}_m \in \overline{\Omega}_c$. Note that our sesquilinear forms defined in Section 1.2 and the finite element construction in Chapter 2 satisfy these requirements. This allows for the parallel assembling of the local operator matrices

$$\underline{A}_l^p = (A_{l,j,m}^p)_{j,m\in\mathcal{I}_l^p}, \qquad \text{where } A_{l,j,m}^p = \sum_{c\in\mathcal{C}_l^p} a_c(\phi_{l,j}, \phi_{l,m}),$$

alltogether they give an additive matrix representation $\underline{A}_l = (\underline{A}_l^1, \ldots, \underline{A}_l^P)$ of the discrete operator A_l

$$A_l\phi_{l,j} = \sum_{p\in\mathcal{P}} \sum_{m\in\mathcal{I}_l^p} A^p_{l,j,m}\phi'_{l,m}.$$

For a consistent vector $\underline{\mathbf{v}}_l \in \underline{\mathbf{X}}_l$ the operation

$$\underline{\mathbf{f}}_l = \underline{A}_l\underline{\mathbf{v}}_l = (\underline{A}_l^p\underline{\mathbf{v}}_l^p)_{p\in\mathcal{P}} \in \underline{\mathbf{X}}'_l$$

gives the additive result without parallel communication. For each process the corresponding local part is obtained by multiplication of the local matrix by the local vector, so the operation is done completely in parallel. Matrix by vector multiplication is the main operation of most numerical algorithms, that is why its efficient realization is the key point for a fast and scalable software implementation.

5.4 Periodic identification

The periodic boundary condition is an *essential boundary condition*, and so it must be satisfied explicitly by finite elements. The periodic boundary condition for a rectangular domain is a simple case, one just needs to identify degrees of freedom on the opposite faces. For a more complex geometry the task is a bit harder, but the approach is the same. We will consider the realization for the domain $\Omega = [0,1]^3$.

A finite element vector $\mathbf{v}_l = \sum_{j\in\mathcal{I}_l} v_{l,j}\phi_{l,j} \in \mathbf{X}_l$ is represented uniquely by its coefficient vector $\tilde{\mathbf{v}}_l = (v_{l,j})_{j\in\mathcal{I}_l}$, with $v_{l,j} = \langle\phi'_{l,j}, \mathbf{v}_l\rangle$. Every index $j \in \mathcal{I}_l$ is associated with just one degree of freedom and the nodal point $\mathbf{n}_j$. The nodal point is located at the mesh element the degree of freedom is based on (vertex, edge, face, cell). If the nodal point $\mathbf{n} = (x, y, z)$ is located at the boundary, then there are one or few identified nodal points calculated by their coordinates:

$$\begin{aligned} I_{\text{per}}(x, y, l) &= (x, y, r), \\ I_{\text{per}}(x, l, z) &= (x, r, z), \\ I_{\text{per}}(l, y, z) &= (r, y, z), \end{aligned}$$

where $(l, r) = (0, 1)$ or $(l, r) = (1, 0)$, few identifications can be applied one after another. The identified nodal points correspond to a set of indices, what defines a set-valued periodic map

$$\theta_l \colon \mathcal{I}_l \to 2^{\mathcal{I}_l} \quad \text{s.t.} \quad \theta_l(j) = \{m \in \mathcal{I}_l \mid I_{\text{per}}(\mathbf{n}_j) = \mathbf{n}_m, \},$$

for convenience assume that $I_{\text{per}}(\mathbf{n}_j) = \mathbf{n}_j$, i.e. $j \in \theta_l(j)$ for all $j \in \mathcal{I}_l$.

Since a degree of freedom $\phi'_{l,j}$ at the boundary is identified with the degrees of freedom $\phi'_{l,m}$, $m \in \theta_l(j)$, and by the definition $v_{l,j} = \langle\phi'_{l,j}, \mathbf{v}_l\rangle$, it follows that $v_{l,j} = v_{l,m}$ for all $m \in \theta_l(j)$. The periodic finite element subspace $\mathbf{X}_l^{\text{per}}$ is defined as follows

$$\mathbf{X}_l^{\text{per}} = \{\mathbf{v}_l \in \mathbf{X}_l \mid v_{l,j} = v_{l,m} \quad \text{for all } j \in \mathcal{I}_l,\ m \in \theta_l(j)\}.$$

A periodic vector $\mathbf{v}_l \in \mathbf{X}_l^{\text{per}}$ has the unique representation in $\mathbb{C}^{\mathcal{N}_l}$ provided by the coefficient vector $\tilde{\mathbf{v}}_l$, the identification is $E_l^{\text{per}} \colon \mathbf{X}_l^{\text{per}} \to \mathbb{C}^{\mathcal{N}_l}$.

Since a basis function $\phi_{l,j}$ corresponding to a boundary mesh element is identified with some other basis functions, we have $\phi_{l,j}^{\text{per}} = \sum_{m\in\theta_l(j)} \phi_{l,m}$, where the sum consists of the original non-periodic basis functions. For a finite element functional $\mathbf{f}_l \in (\mathbf{X}_l^{\text{per}})'$ we have a non-unique additive representation in $\mathbb{C}^{\mathcal{N}_l}$ provided by the coefficient vector $\tilde{\mathbf{f}}_l$ with

$$\langle\mathbf{f}_l, \phi_{l,j}^{\text{per}}\rangle = \sum_{m\in\theta_l(j)} \langle\mathbf{f}_l, \phi_{l,m}\rangle = \sum_{m\in\theta_l(j)} f_{l,m} = f_{l,j}^{\text{per}}.$$

With respect to the standard inner product in $\mathbb{C}^{\mathcal{N}_l}$ we obtain the adjoint operator for E_l^{per}, $(E_l^{\text{per}})' \colon \mathbb{C}^{\mathcal{N}_l} \to (X_l^{\text{per}})'$ s.t. $(E_l^{\text{per}})'(\tilde{\mathbf{f}}_l) = \mathbf{f}_l$, this mapping is defined by

$$\mathbf{f}_l = \sum_{j \in \mathcal{I}_l} \sum_{m \in \theta_l(j)} f_{l,m} \phi'_{l,m},$$

the mapping is not one-to-one. As a norm in X_l^{per} and $(X_l^{\text{per}})'$ we define the standard norm in $\mathbb{C}^{\mathcal{N}_l}$.

Having a linear ordering $<$ for nodal points $\mathbf{n}_j, \mathbf{n}_m$, we define a subspace of $\mathbb{C}^{\mathcal{N}_l}$

$$\hat{\mathbf{F}}_l = \{\tilde{\mathbf{f}}_l \in \mathbb{C}^{\mathcal{N}_l} \mid f_{l,j} = 0 \text{ if } \mathbf{n}_j \neq \min_{m \in \theta_l(j)} \mathbf{n}_m\},$$

which gives the unique representation of functionals $\mathbf{f}_l \in (\mathbf{X}_l^{\text{per}})'$, $(E_l^{\text{per}})'$ restricted to $\hat{\mathbf{F}}_l$ is one-to-one. The smallest nodal point among the identified ones is the master point, only the corresponding index holds data.

As we see the realization of the periodic boundary condition has a lot of common with the realization of the parallel finite elements, both are based on an identification of degrees of freedom. The vectors are represented as the consistent (periodic) vectors with a few equal coordinates, while the functionals have the additive representation. The only significant difference is that in the parallel case the vectors and the functionals are distributed. The periodic identification can be realized on top of the parallel distribution.

5.5 Parallel linear algebra

In numerics most of algorithms are formulated in terms of linear algebra: matrix-vector product, scalar product, linear combination of vectors, etc. If we manage to define these elementary operations for the parallel finite elements correctly, then we can be sure that complex algorithms are also correct. In fact, a reliable and efficient implementation of the parallel linear algebra is based on a very simple concept, we just use consistent vectors in $\underline{\mathbf{X}}_l$ as a representation of $\mathbf{X}_l$ and additive vectors in $\underline{\mathbf{X}}'_l$ as a representation of $\mathbf{X}'_l$. This automatically yields correct parallel algorithms, if the following operations are included:

a) A unique additive parallel representation is obtained by collect: $\underline{\mathbf{X}}'_l \to \hat{\underline{\mathbf{X}}}'_l$, defined by modification of vector elements, for all $p \in \mathcal{P}$, $j \in \mathcal{I}_l^p$

$$\text{collect}(f_{l,j}^p) = \begin{cases} \sum_{q \in \pi_l(j)} f_{l,j}^q, & p = \min \pi_l(j), \\ 0, & \text{else}. \end{cases}$$

b) A unique additive periodic representation is obtained by $\text{collect}^{\text{per}}$: $\hat{\underline{\mathbf{X}}}'_l \to \hat{\underline{\mathbf{X}}}'_l \cap \hat{\mathbf{F}}_l$, defined by modification of vector elements, for all $p \in \mathcal{P}$, $j \in \hat{\mathcal{I}}_l^p$

$$\text{collect}^{\text{per}}(f_{l,j}^p) = \begin{cases} \sum_{m \in \theta_l(j)} f_{l,m}^{\min \pi_l(m)}, & \mathbf{n}_j = \min_{m \in \theta_l(j)} \mathbf{n}_m, \\ 0, & \text{else}. \end{cases}$$

c) A consistent result of local corrections is obtained by accumulate: $\underline{\mathbf{X}}_l^{\mathcal{P}} \to \underline{\mathbf{X}}_l$, defined by modification of vector elements, for all $p \in \mathcal{P}$, $j \in \mathcal{I}_l^p$

$$\text{accumulate}(v_{l,j}^p) = \sum_{q \in \pi_l(j)} v_{l,j}^q.$$

d) A periodic result of local corrections is obtained by $\text{accumulate}^{\text{per}}\colon \underline{\mathbf{X}}_l \to \underline{\mathbf{X}}_l \cap \mathbf{X}_l^{\text{per}}$, defined by modification of vector elements, for all $p \in \mathcal{P}$, $j \in \hat{\mathcal{I}}_l^p$

$$\text{accumulate}^{\text{per}}(v_{l,j}^p) = \sum_{m \in \theta_l(j)} v_{l,m}^{\min \pi_l(m)}.$$

For each process the operations a - d require communication only with limited number of other processes, the neighboring processes exchange data (for the parallel collect/accumulate) and processes having a boundary region exchange data (for the periodic collect/accumulate). The size of the transmitted data is small, just a fraction of the local vector. Global communication is required only for the inner products and the norms. This results in a fast and scalable parallel implementation.

The data exchange in the collect/accumulate routines is done by identifying the indices via the position of the nodal points, no global numbering is required. A process p sends to a process q the following message: size of the message s, nodal points $(\mathbf{n}_{j_1}, \dots, \mathbf{n}_{j_s})$ s.t. $q \in \pi_l(j_m)$, $m = 1, \dots, s$, the corresponding elements of a local vector $(v_{l,j_1}^p, \dots, v_{l,j_s}^p)$. At the same time the process p receives the similar message from the process q. After receiving the local indices are determined using the efficient hash map container $\mathcal{N}_l^p$.

Although such a messaging is very flexible, it requires to pass extra data (the nodal points). In practice one can implement a cache s.t. for common messages processes send only a special identificator instead of the nodal points array. It is possible because for the same finite elements and the parallel distribution every time processes send the same nodal points.

We distinguish six types of parallel operators. Most of numerical algorithms and all which we described consist of these few elements.

Discrete operators $A_l\colon \mathbf{X}_l \to \mathbf{X}_l'$ are represented additively by parallel distributed matrices $\underline{A}_l\colon \underline{\mathbf{X}}_l \to \hat{\underline{\mathbf{X}}}_l' \cap \hat{\mathbf{F}}_l$, the operation is defined by

$$\underline{A}_l\underline{\mathbf{v}}_l = \text{collect}^{\text{per}}\Big(\text{collect}(\underline{A}_l^p\underline{\mathbf{v}}_l^p)\Big),$$

where $\underline{A}_l^p$ is represented locally as a sparse matrix, the matrix-vector products $\underline{A}_l^p\underline{\mathbf{v}}_l^p$ are computed in parallel.

Preconditioners $T_l\colon \mathbf{X}_l' \to \mathbf{X}_l$ are represented by parallel distributed matrices $\underline{T}_l\colon \underline{\mathbf{X}}_l' \to \underline{\mathbf{X}}_l \cap \mathbf{X}_l^{\text{per}}$, the operation is defined by

$$\underline{T}_l\underline{\mathbf{f}}_l = \text{accumulate}^{\text{per}}\Big(\text{accumulate}(\underline{T}_l^p\underline{\mathbf{f}}_l^p)\Big),$$

where $\underline{T}_l^p$ is a sparse matrix or even an operation defined as a routine, application of the local preconditioners happens in parallel.

The *prolongation* $I_{l-1,l}\colon \mathbf{X}_{l-1} \to \mathbf{X}_l$ is represented by parallel distributed matrices $\underline{I}_{l-1,l}\colon \underline{\mathbf{X}}_{l-1} \to \underline{\mathbf{X}}_l$, the operation is defined by

$$\underline{I}_{l-1,l}\underline{\mathbf{v}}_{l-1} = (\underline{I}_{l-1,l}^p\underline{\mathbf{v}}_{l-1}^p),$$

where $\underline{I}_{l-1,l}^p$ is a sparse matrix. The operation is pure local, no communication is required.

The *restriction* $R_{l,l-1} = I_{l-1,l}'\colon \mathbf{X}_l' \to \mathbf{X}_{l-1}'$ is represented additively by parallel distributed matrices $\underline{R}_{l,l-1}\colon \underline{\mathbf{X}}_l' \to \hat{\underline{\mathbf{X}}}_l' \cap \hat{\mathbf{F}}_l$, the operation is defined by

$$\underline{R}_{l,l-1}\underline{\mathbf{f}}_l = \text{collect}^{\text{per}}\Big(\text{collect}(\underline{R}_{l,l-1}^p\underline{\mathbf{f}}_l^p)\Big),$$

where $\underline{R}_{l,l-1}^p$ is a sparse matrix.

As described in Section 5.2, the *parallel dual pairing* $(\cdot,\cdot)\colon \underline{\mathbf{X}}_l' \times \underline{\mathbf{X}}_l \to \mathbb{C}$ is defined by the local inner products

$$(\underline{\mathbf{f}}_l, \underline{\mathbf{v}}_l) = \sum_{p \in \mathcal{P}} (\underline{\mathbf{f}}_l^p, \underline{\mathbf{v}}_l^p).$$

It requires to take a sum of the local contributions, so global communication is needed. The norm $\|\cdot\|_{\underline{\mathbf{X}}_l'}: \underline{\mathbf{X}}_l' \to \mathbb{R}$ is defined by

$$\|\underline{\mathbf{f}}_l\|_{\underline{\mathbf{X}}_l'} = (\hat{\underline{\mathbf{f}}}_l, \hat{\underline{\mathbf{f}}}_l), \quad \text{where } \hat{\underline{\mathbf{f}}}_l = \text{collect}^{\text{per}}\Big(\text{collect}(\underline{\mathbf{f}}_l)\Big).$$

A *linear combination* of vectors in $\underline{\mathbf{X}}_l$ or $\underline{\mathbf{X}}_l'$ is realized by the corresponding linear combination of their local parts. It is a pure local operation and so no communication is required.

5.6 Scalable interprocess communication

The most challenging aspect of parallel computing is scalability. Without a good scalable design it is not possible to leverage full power of modern parallel clusters consisting of thousands processors. The parallel model described in this chapter provides a high-level scalable design based on few types of parallel operations. But the final performance also depends on a low-level MPI realization. Here we want to share some thoughts.

The current MPI standard 2.2 introduce two types of interprocess communications: point-to-point communication and collective communication. Both can be used in an implementation of our parallel model, but they result in different scalability. To analyze this we need to consider parallel communication patterns arising in our parallel model.

For a cubic domain imagine a parallel distribution based on the regular grid, so the cube consists of many equal subcubes, each one belongs to a different process. In our model each cube shares boundary degrees of freedom with neighbors, so an inner cube has to communicate with 26 other cubes, 6 of them share a face (much data), 12 share an edge (less data) and 8 share a vertex (few or no data if there are no vertex-based degrees of freedom). There are two typical realizations of parallel communication:

1. **Few simultaneous nonblocking point-to-point communications**. Every process initiates nonblocking `MPI_Isend` and `MPI_Irecv` (or `MPI_Get` and `MPI_Put`) to all neighbors, the communications run simultaneously. If P is the number of processes, then the total number of simultaneous communications is $2 \cdot 26P = 52P$. So many communications put a high load on memory bandwidth and the interconnect, that results in not optimal performance. The only advantage of this implementation is the possibility to use computations/communications overlapping for a slow interconnect.

2. **One collective blocking communication**. All processes call `MPI_Alltoallv` which executes the parallel exchange. Not neighboring processes just specify zero datasize to send/receive, so the communication occurs only for neighboring processes. The exact behavior of `MPI_Alltoallv` depends on a given MPI implementation, modern implementations provide that the operation is non-synchronizing, i.e. a process goes on once it gets/sends all data, while other processes still keep exchanging.

Although the variant 2 is already sufficiently good it has a lack, no mechanism is provided against improper load balancing and lost of synchronization. The following algorithm is intended to correct the limitation.

Let us introduce a graph of the parallel distribution, which is represented by the matrix $G = (G_{i,j})$, $i, j = 1, \dots, P$. If a processes i and j have a common interface, then $G_{i,j} = 1$, else $G_{i,j} = 0$. By the definition, the matrix G is symmetric and, in our case, it is sparse. The matrix can be constructed during the parallel distribution. Next we convert the graph G into a weighted graph, just enumerate anyway all links of each vertex. Then the graph should be copied to all processes. Our goal is to define an optimal order of point-to-point communications between the processes, that is done in three steps.

1. **Initialization**. The weights of the original graph denote the order of communications, a process i first initiates `MPI_Isend` (nonblocking) and `MPI_Recv` (blocking) to the process j s.t. $G_{i,j} = 1$, waits until the sending is complete, then do the same with the process l s.t. $G_{i,l} = 2$, and so on. The communications happen one after another, so the total number of simultaneous communications is $2P$, while the interconnect is fully loaded. The first few parallel communications occur in this order.

2. **Measure computational time**. Some processes might get more work, than the others or some nodes compute a bit slower, that results in different computational time for processes. The computational time for local work is measured and then is sent to all processes by `MPI_Allgather`. This information is a basis for scheduling the communications.

3. **Scheduling**. The idea is very simple, processes should first communicate with fast processes. Let i_k be a process numbering in the time increasing order, now change the communication order by renumbering links of the graph G. We start with the vertex i_1 and enumerate its links also in the time increasing order, then do the same with the vertex i_2 and so on.

This algorithm ensures that a fast process does not spend time waiting for a slow one. The steps 2-3 need not be done everytime, just as often as it is appropriate. Most MPI implementations (see, e.g. [37]) use blocking consecutive point-to-point operations in the `MPI_Alltoallv` algorithm (except for very small messages), so our method is at least as good, but it provides higher performance for large-scale clusters suffering the synchronization problem.

Chapter 6

Numerical results

Here we present and discuss results obtained in numerical experiments with our Maxwell code. In order to simplify further explanation we first introduce material configurations and other parameters which will be used in this chapter very often.

In practice a material distribution of the photonic crystal usually consists of only two materials, air (or vacuum) and a dielectric material, so the electric permittivity ε may have only two values. We will consider the distributions made of air $\varepsilon = 1$ and silicon $\varepsilon = 13$. In terms of the refractive index it equals to 1 and 3.6 correspondingly. Our fundamental domain is the unit cube $\Omega = [0, 1]^3$, a configuration is defined there, multiple copies of the domain build the crystal structure.

We consider three material distributions.

Configuration 0. $\varepsilon = 1$, there is no material, only air (vacuum). It is a test configuration. We use it to check the code and provide some exact errors.

Configuration 1. The distribution is called "scaffold" structure, it is presented in Figure 6.1. The structure is very simple and highly symmetric, it consists of a silicon frame and air, the frame thickness is 0.125 in the unit cube and 0.25 in the periodic structure.

Configuration 2. The distribution is a layered structure, it is shown in Figure 6.2. The structure is highly symmetric in xy plane, but is not symmetric in z direction. The configuration consists of silicon blocks and air, the block thickness in the periodic structure is 0.25.

The configurations 1 and 2 are described in [12], it is known that they exhibit some band gaps. We take these distributions for the sake of convenience, it allows to compare our results with published data.

In numerical computations we use two pairs of finite elements, every pair consists of $\mathbf{X}_{h,\mathbf{k}}$ and $Q_{h,\mathbf{k}}$ elements.

Elements LOhex. $\mathbf{X}_{h,\mathbf{k}}$ is the lowest order Nédélec elements and $Q_{h,\mathbf{k}}$ is the linear elements (trilinear $\mathcal{Q}^{1,1,1}$) on hexahedra, recall Section 2.1.

Elements HOhex. $\mathbf{X}_{h,\mathbf{k}}$ is the $\mathbf{H}(\text{curl})$-conforming hierarchical higher order ($p = p_e = p_f = p_c = 1$) elements and $Q_{h,\mathbf{k}}$ is the H^1-conforming hierarchical higher order ($p = p_e = p_f = p_c = 2$) elements on hexahedra, recall Section 2.2.

Remember that we use $\mathbf{H}(\text{curl})$-conforming elements $\mathbf{X}_{h,\mathbf{k}}$ in the main eigenvalue Problem 3.1 and its preconditioning Problem 4.1, H^1-conforming elements $Q_{h,\mathbf{k}}$ are used in the projection $P_{h,\mathbf{k}}$ (Section 3.2) and the hybrid smoother (Section 4.3).

Tables with numerical results provided in this chapter usually depend on the refinement level. For simplicity we describe all details of the refinement levels only once and later just

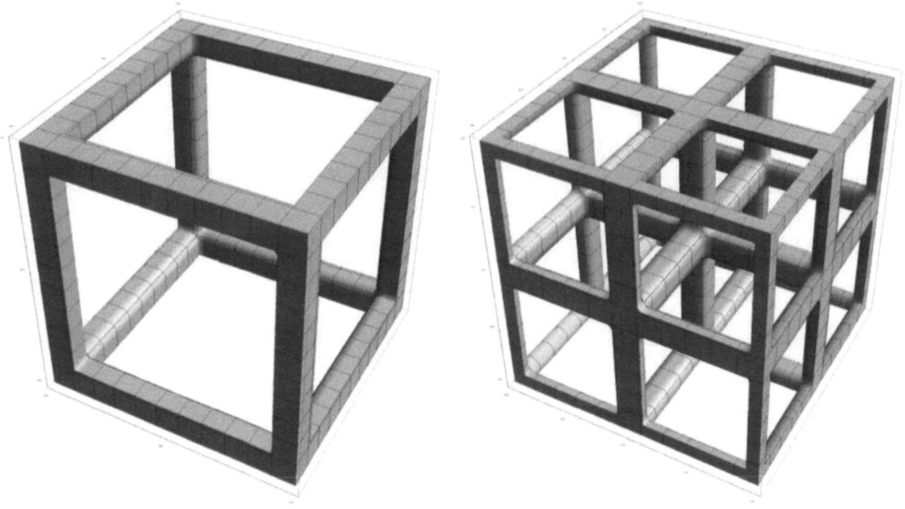

Figure 6.1: Configuration 1, the scaffold structure in the domain $[0,1]^3$ (left) and $[0,2]^3$ (right).

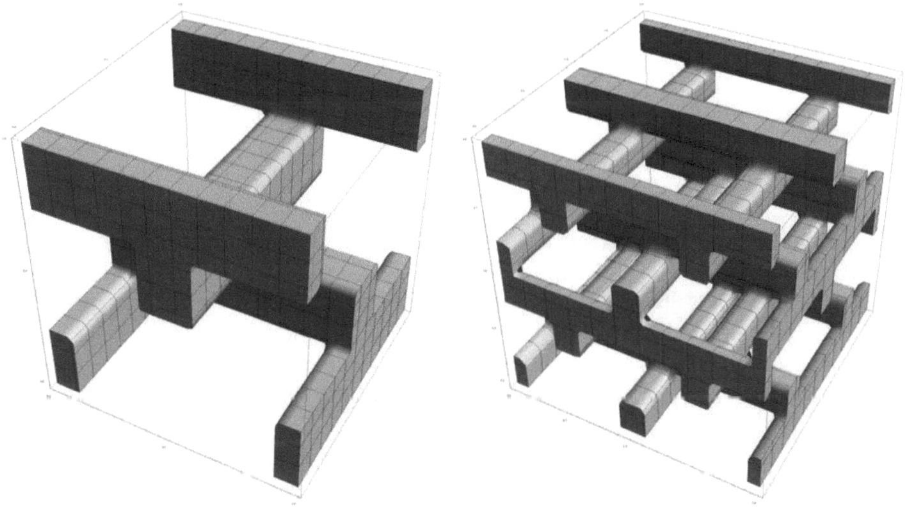

Figure 6.2: Configuration 2, the layered structure in the domain $[0,1]^3$ (left) and $[0,2]^3$ (right).

		Lowest order (LOhex)		Higher order (HOhex)	
Level	Cells	DoFs in $\mathbf{X}_h$	DoFs in Q_h	DoFs in $\mathbf{X}_h$	DoFs in Q_h
2	64			1,944	729
3	512	1,944	729	13,872	4,913
4	4,096	13,872	4,913	104,544	35,937
5	32,768	104,544	35,937	811,200	274,625
6	262,144	811,200	274,625	6,390,144	2,146,689
7	2,097,152	6,390,144	2,146,689		
8	16,777,216	50,725,632	16,974,593		

Table 6.1: Refinement levels of Elements LOhex and HOhex, number of cells and degrees of freedom.

mention their numbers. The level 0 is one hexahedron, then the regular refinement produces 8 times more cells on every next level. Table 6.1 contains statistics for Elements LOhex and HOhex.

The level 3 is the minimal level where Configurations 1-2 can be represented exactly, i.e. the material distributions are aligned with cells.

6.1 Analytical solutions

Let us consider the original problem resulted from the Maxwell equations and the Bloch ansatz, also assume $\varepsilon = 1$.

$$(\nabla + \mathrm{i}\mathbf{k}) \times (\nabla + \mathrm{i}\mathbf{k}) \times \mathbf{u} = \lambda \mathbf{u} \qquad \text{in } \Omega, \tag{6.1}$$

$$(\nabla + \mathrm{i}\mathbf{k}) \cdot \mathbf{u} = \mathbf{0} \qquad \text{in } \Omega, \tag{6.2}$$

$$\mathbf{u} \text{ is periodic on } \partial\Omega. \tag{6.3}$$

For a fixed vector $\mathbf{k}$ there exist constant vectors $\mathbf{k}_1^\perp$ and $\mathbf{k}_2^\perp$ s.t.

$$|\mathbf{k}_1^\perp| = |\mathbf{k}_2^\perp| = 1,$$
$$\mathbf{k} \times \mathbf{k}_1^\perp = |\mathbf{k}|\mathbf{k}_2^\perp,$$
$$\mathbf{k} \times \mathbf{k}_2^\perp = -|\mathbf{k}|\mathbf{k}_1^\perp,$$

so vectors $\mathbf{k}, \mathbf{k}_1^\perp, \mathbf{k}_2^\perp$ build an orthogonal basis. First, one may see that $\mathbf{k}_1^\perp$ and $\mathbf{k}_2^\perp$ satisfy (6.2) and (6.3) since they are constants and $\mathbf{k} \cdot \mathbf{k}_j^\perp = 0$. Second, if one put e.g. $\mathbf{u} = \mathbf{k}_1^\perp$ in (6.1) it gives

$$(\nabla + \mathrm{i}\mathbf{k}) \times (\nabla + \mathrm{i}\mathbf{k}) \times \mathbf{k}_1^\perp = (\nabla + \mathrm{i}\mathbf{k}) \times (\mathrm{i}|\mathbf{k}|\mathbf{k}_2^\perp) = -\mathrm{i}^2|\mathbf{k}|^2\mathbf{k}_1^\perp.$$

So we conclude that $\mathbf{k}_1^\perp$ and $\mathbf{k}_2^\perp$ are the eigenfunctions with eigenvalues $|\mathbf{k}|^2$. These solutions will be used to check numerical results.

6.2 Performance of the eigenvalue solver

The performance of the LOBPCG eigenvalue solver depends on many variables. First, it depends on the spectrum itself, the more clustered (closer to each other) eigenvalues are, the more iterations it takes to compute them.

Second, the performance is affected by quality of the preconditioning, i.e. the accuracy up to which we solve the auxiliary linear Maxwell problem by an iterative solver. Of course, a higher preconditioner accuracy improves the eigenvalue solver, but there is a certain threshold after

	Configuration 0		Configuration 2	
δ	LOBPCGit	Prec.conv.	LOBPCGit	Prec.conv.
10^{-5}	11	0.316	12	0.356
10^{-4}	11	0.317	12	0.338
10^{-3}	10	0.277	12	0.356
0.01	10	0.278	12	0.357
0.1	10	0.277	12	0.360
1	11	0.267	12	0.349
10	11	0.264	16	0.353
100	21	0.184	39	0.346

Table 6.2: Influence of the parameter δ on iteration number of the eigenvalue solver. It is also shown the convergence rate of the Maxwell multigrid (the eigensolver preconditioner).

that increasing the accuracy does not reduce the number of eigenvalue iterations considerably. Since extra preconditioning takes more time there is some optimal point for the precision. In our experiments the Maxwell residual reduction ϵ_T is set to 0.01.

Third, stability and performance of the eigenvalue solver depends on accuracy of the projection to the divergence-free fields. On the projection step we solve the internal linear Laplace problem inexactly by an iterative solver. This accuracy is a very important and the most tricky parameter. If the accuracy is not good enough then the eigensolver may take more iterations or even converge to zero. Unfortunately in our case there is no theoretical estimation for the optimal precision, so one should try and figure it out for every single case. Usually it is a good idea to start with some high precision and then decrease it up to a point when the convergence is still stable and the number of iterations does not grow. It is affected by many other parameters, e.g. required eigenvalue precision and mesh size. In our computations the Laplace residual reduction ϵ_P is set to 10^{-8}.

Fourth, the convergence rate of the eigenvalue solver depends on the regularization parameter δ (see Section 3.4 and Problem 4.1). We do not have a theory for that. The parameter affects convergence rate of the linear Maxwell solver, the larger δ the more well-conditioned matrix A^δ is the better convergence rate of the iterative solver. But as we see from the computations, a large δ is bad for the eigenvalue solver convergence. Table 6.2 shows the influence of δ on the eigenvalue solver performance. Here we compute the first 5 eigenvalues in a block of 10 for $\mathbf{k} = (3, 1, -2)$ using Elements LOhex on level 5, the eigensolver accuracy is $\epsilon_E = 10^{-4}$, 4 iterations of the preconditioner (the linear Maxwell solver) are performed. In the notations from Section 3.4 the parameters are: $n = 5$, the eigensolver stopping criterion is $\|\mathbf{r}_j\|_{M^{-1}} \leq \epsilon_E$. As we see in the table, while the preconditioner improves convergence, the eigensolver convergence could degrade. There is some optimal point (interval) for δ, but what is more important is that δ should not be larger than a threshold, otherwise the iteration count grows significantly. We noticed that in general δ should not be larger than the first eigenvalue λ_1 and not be too small, the choice $\frac{\lambda_1}{10}$ is suboptimal.

We observe that number of the eigenvalue iterations is almost independent of mesh size. A typical behavior is shown in Table 6.3. The table shows the iteration count when there are m converged eigenvectors ($m = 1, \ldots, 10$), they converge one by one during one run. We iterate a block of 15 vectors for Configuration 2, $\mathbf{k} = (3, 1, -2)$ using Elements LOhex, $\epsilon_E = 10^{-4}$, $\epsilon_T = 0.001$ (extra accuracy is given in order to minimize influence of the preconditioning). Every next refinement level gives 8 times more degrees of freedom, but as we see the difference in the iteration numbers is insignificant. Although the table is given for only one $\mathbf{k}$, Configuration and Elements this behavior is common for all other parameters.

Eigenvalue #	Level 3	Level 4	Level 5	Level 6
1	6	6	6	7
2	6	7	7	7
3	7	7	7	8
4	7	8	8	8
5	9	10	10	11
6	9	10	11	11
7	11	12	12	12
8	11	12	12	12
9	12	13	13	13
10	12	13	14	14

Table 6.3: Number of iteration the eigensolver takes to converge for the first m eigenvalues depending on refinement level.

6.3 Parallel performance

In order to compute a complex material distribution or/and with high accuracy and speed one needs an efficient parallel implementation. A MPI-based distributed memory parallel implementation can run on a huge computing cluster which provides computing power and memory resources unavailable on a single workstation.

All computations have been performed on the parallel cluster IC1 in the Scientific Supercomputing Center of University Karlsruhe. The cluster consists of 200 8-way Intel Xeon X5355 nodes. Each of these nodes contains 16 GB of main memory and two Quad-core Intel Xeon processors which run at a clock speed of 2.667 GHz and have 2x4 MB L2 cache. An important component of the cluster is the InfiniBand 4X DDR interconnect. All nodes are attached to this interconnect which is characterized by its very low latency of below 2 microseconds and a point to point bandwidth between two nodes of more than 1300 MB/s. At the moment, this architecture is typical for HPC clusters.

One of the most important aspects of a parallel code is the parallel scalability, it may be defined as $\frac{t(P_0)}{t(P_1)}$, where $t(P_j)$ is the computational time on P_j processors. In an ideal case the scalability should be close to $\frac{P_1}{P_0}$, but due to some practical limitations it is usually in between 1 and $\frac{P_1}{P_0}$, where 1 means no performance profit.

In Table 6.4 we demonstrate parallel performance of our implementation. Here we compute the first 10 eigenvalues in a block of 15 for Configuration 2 and $\mathbf{k} = (3, 1, -2)$ using Elements LOhex, $\epsilon_E = 10^{-4}$, refinement level 7.

We observe a good scalability up to 256 processors, a slow execution on 512 processors will be explained later. As a multigrid convergence rate in the table we denote the convergence rate of the Maxwell linear iterative solver with the multigrid preconditioner, which by-turn is used as a preconditioner for the eigenvalue solver. One may see that the convergence rate does not deteriorate considerably when the processor number increases. It is a good result since our multigrid smoother is a block-Jacobi smoother resulting from the parallel domain decomposition, so it becomes worse when processor number increases. With a good convergence rate it is enough just few iterations (we do 2-8) of the linear solver for the eigenvalue solver preconditioning.

In Table 6.5 we indicate influence of dataset size on performance. We keep the same settings as before, but fix the processor number to 256 and vary mesh size (refinement level). Every refinement level increases number of DoFs in 8 times, so in an ideal case the computational time should also change in 8 times. One may notice that it does not happen. The reason is that a MPI-based parallelization is inefficient when the dataset is too small, it happens because

Processors	Multigrid conv. rate	Total time, min	Scalability
64	0.28	37:47	
			2.12
128	0.28	17:45	
			1.80
256	0.27	9:54	
			1.19
512	0.29	8:20	

Table 6.4: Parallel scalability and Maxwell multigrid convergence on the refinement level 7.

Refinement level	DoFs	Total time, min	Speed factor
5	104,544	1:05	
			1.78
6	811,200	1:56	
			5.12
7	6,390,144	9:54	
			8.09
8	50,725,632	80:21	

Table 6.5: Parallel performance on 256 processors.

interprocessor communications take more time than computations. That is why a smaller problem may take disproportionately less time than a larger problem. This fact may explain the bad scalability on 512 processors observed in Table 6.4, increasing the number of processors we made the dataset size per processor too small.

The same parallel performance trend can be shown for other $\mathbf{k}$, Configurations and Elements.

6.4 Eigenvalue precision

In this section we study the precision of the eigenvalue computations obtained for different finite elements and Configurations. For a fixed $\mathbf{k}$ we compute the eigenvalues on successive refinement levels, so one can observe the h-convergence of the finite elements.

We compute the first 10 eigenvalues for $\mathbf{k} = (3, 1, -2)$ up to the precision $\epsilon_E = 10^{-4}$. Let $\lambda_j(l_m)$ be an eigenvalue of number j on the refinement level l_m, define

$$\Delta_j(l_m, l_{m+1}) = |\lambda_j(l_m) - \lambda_j(l_{m+1})|,$$
$$\gamma_j(l_m) = \log_2\left(\frac{\lambda_j(l_{m-1}, l_m)}{\lambda_j(l_m, l_{m+1})}\right).$$

$\gamma_j(l_m)$ characterizes a practical h-convergence order, so it is the most interesting parameter. For convenience we provide separate "h-convergence" tables consisting of $\lambda_j(l_m, l_{m+1})$ and $\gamma_j(l_m)$.

Let us start with our benchmark problem for Configuration 0. In Tables 6.6-6.7 we observe eigenvalue convergence. From Section 6.1 we remember that the first two analytical eigenvalues are $|\mathbf{k}|^2 = 3^2 + 1^2 + 2^2 = 14$, so our numerical eigenvalues converge to the correct values. We checked that the eigenvectors are correct too. Since $\varepsilon = 1$ we have high regularity of the original problem, so the practical convergence order $\gamma_j(l_m)$ is 2, it is the maximal order allowed by Theorem 1.6 for the lowest order Nédélec elements. Now consider Tables 6.8-6.9. One

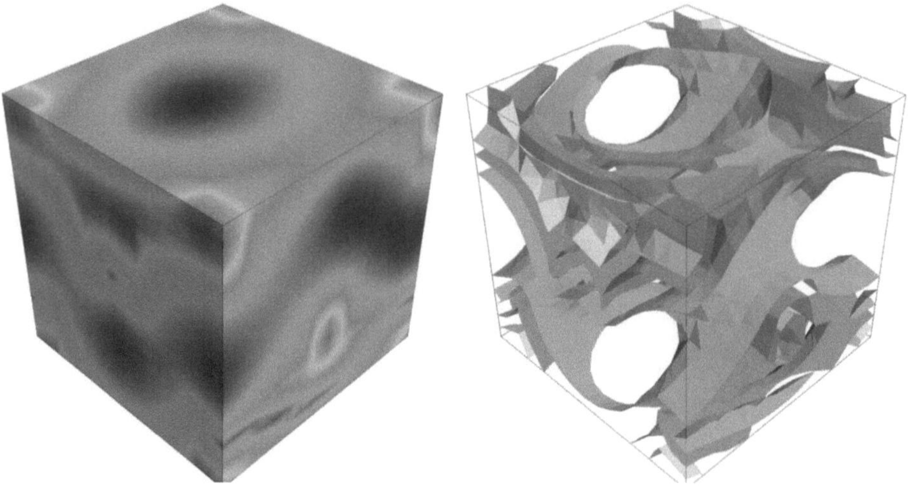

Figure 6.3: Eigenfunction #5 of Configuration 2. Amplitude of the **H** field, surface plot (left) and few color-coded isosurfaces (right).

may see convergence to the same eigenvalues as in the former case, but the convergence order is much better. Thanks to a better approximation order of the higher order elements the practical convergence order $\gamma_j(l_m)$ is 4.

Although the first two analytical solutions are just constant vectors one can notice that the numerical eigenvalues are not equal to 14, they just converge to that. In classical finite elements it does not happen, because constants can be represented exactly, and so both solutions coincide. In our case of the k-modified finite elements a constant vector cannot be represented exactly in the finite element spaces.

Now let us go to a more complicated case of Configurations 1 and 2. ε is a piecewise constant function, so regularity of the problem is restricted. Tables 6.6-6.7 contain the data for Configuration 1 and 2 solved by Elements LOhex. We notice that the practical convergence order is below 2, it is the result of the lower regularity. A visualization of an eigenfunction is shown at Figure 6.3.

What happens with Elements HOhex in Configuration 2 is shown in Tables 6.10-6.11. First, a positive point is that they converge to the same eigenvalues as Elements LOhex. A negative point is that the practical convergence order is still below 2, much lower than 4 which the elements could provide. Of course, this result is not new in numerics. The convergence order is always limited by regularity of a solution. The only way to address the problem is to use an adaptivity concept. However, even in our case the higher order elements are not completely useless. Comparing Table 6.6 and Table 6.10 one may notice that Elements HOhex on level l give the same accuracy as Elements LOhex on level $l+2$. This results in 8 times less DoFs and the sparse matrices contain 2.5 times less entries. For instance, Elements HOhex on level 4 give $104,544$ DoFs and the matrix A has $10,328,256$ entries, while Elements LOhex on level 6 give $811,200$ DoFs and $26,124,480$ entries. From performance point of view it can be advantageous to use the higher order elements even without adaptivity.

Configuration 0						
Eigenvalue #	level 3	level 4	level 5	level 6	level 7	level 8
1	14.12814	14.03193	14.00798	14.00199	14.00050	14.00012
2	14.12814	14.03193	14.00798	14.00199	14.00050	14.00012
3	15.95361	15.82272	15.79015	15.78202	15.77998	15.77948
4	15.95361	15.82272	15.79015	15.78202	15.77998	15.77948
5	28.89523	28.48222	28.37976	28.35419	28.34781	28.34621
6	28.89523	28.48222	28.37976	28.35419	28.34781	28.34621
7	30.72071	30.27300	30.16193	30.13422	30.12729	30.12556
8	30.72071	30.27300	30.16193	30.13422	30.12729	30.12556
9	42.06735	41.19817	40.98340	40.92987	40.91650	40.91316
10	42.06735	41.19817	40.98340	40.92987	40.91650	40.91316

Configuration 1						
Eigenvalue #	level 3	level 4	level 5	level 6	level 7	level 8
1	4.00678	3.97226	3.95812	3.95272	3.95070	3.94994
2	4.76732	4.73162	4.71647	4.71062	4.70841	4.70759
3	9.37758	8.81912	8.66063	8.61324	8.59845	8.59366
4	10.62938	9.88117	9.67155	9.60867	9.58886	9.58238
5	12.03825	11.32434	11.08231	11.00678	10.98276	10.97488
6	12.81253	11.84988	11.62060	11.55728	11.53851	11.53267
7	13.12710	12.23778	12.02210	11.96278	11.94539	11.94005
8	16.56371	15.32280	15.01242	14.92683	14.90184	14.89421
9	16.86278	15.83867	15.55781	15.48168	15.46060	15.45457
10	17.79952	16.66801	16.37232	16.29261	16.27043	16.26405

Configuration 2						
Eigenvalue #	level 3	level 4	level 5	level 6	level 7	level 8
1	3.86613	3.81920	3.80119	3.79449	3.79202	3.79111
2	4.13949	4.09004	4.07099	4.06392	4.06132	4.06036
3	5.01881	4.87990	4.82904	4.81021	4.80322	4.80063
4	5.38737	5.24131	5.18723	5.16714	5.15967	5.15690
5	10.19397	9.68279	9.54135	9.50006	9.48747	9.48348
6	10.46622	9.95338	9.80933	9.76693	9.75392	9.74977
7	12.22023	11.26907	11.02740	10.95948	10.93931	10.93307
8	12.35505	11.40819	11.16670	11.09883	11.07870	11.07247
9	13.78493	12.78181	12.51608	12.43920	12.41575	12.40829
10	13.95290	12.93797	12.66950	12.59189	12.56823	12.56071

Table 6.6: The eigenvalues computed with Elements LOhex.

Configuration 0									
j	$\Delta_j(3,4)$	$\gamma_j(4)$	$\Delta_j(4,5)$	$\gamma_j(5)$	$\Delta_j(5,6)$	$\gamma_j(6)$	$\Delta_j(6,7)$	$\gamma_j(7)$	$\Delta_j(7,8)$
1	0.09620	2.01	0.02396	2.00	0.00598	2.00	0.00150	2.00	0.00037
2	0.09620	2.01	0.02396	2.00	0.00598	2.00	0.00150	2.00	0.00037
3	0.13089	2.01	0.03257	2.00	0.00813	2.00	0.00203	2.00	0.00051
4	0.13089	2.01	0.03257	2.00	0.00813	2.00	0.00203	2.00	0.00051
5	0.41301	2.01	0.10246	2.00	0.02556	2.00	0.00639	2.00	0.00160
6	0.41301	2.01	0.10246	2.00	0.02556	2.00	0.00639	2.00	0.00160
7	0.44770	2.01	0.11107	2.00	0.02771	2.00	0.00692	2.00	0.00173
8	0.44770	2.01	0.11107	2.00	0.02771	2.00	0.00692	2.00	0.00173
9	0.86917	2.02	0.21477	2.00	0.05353	2.00	0.01337	2.00	0.00334
10	0.86917	2.02	0.21477	2.00	0.05353	2.00	0.01337	2.00	0.00334
Configuration 1									
j	$\Delta_j(3,4)$	$\gamma_j(4)$	$\Delta_j(4,5)$	$\gamma_j(5)$	$\Delta_j(5,6)$	$\gamma_j(6)$	$\Delta_j(6,7)$	$\gamma_j(7)$	$\Delta_j(7,8)$
1	0.03452	1.29	0.01414	1.39	0.00540	1.42	0.00202	1.43	0.00075
2	0.03570	1.24	0.01515	1.37	0.00585	1.41	0.00221	1.42	0.00082
3	0.55846	1.82	0.15850	1.74	0.04738	1.68	0.01479	1.63	0.00479
4	0.74821	1.84	0.20962	1.74	0.06289	1.67	0.01981	1.61	0.00648
5	0.71391	1.56	0.24202	1.68	0.07553	1.65	0.02402	1.61	0.00787
6	0.96265	2.07	0.22928	1.86	0.06332	1.75	0.01877	1.69	0.00583
7	0.88932	2.04	0.21568	1.86	0.05932	1.77	0.01739	1.70	0.00534
8	1.24091	2.00	0.31039	1.86	0.08559	1.78	0.02499	1.71	0.00763
9	1.02410	1.87	0.28086	1.88	0.07613	1.85	0.02109	1.81	0.00602
10	1.13151	1.94	0.29568	1.89	0.07971	1.85	0.02218	1.80	0.00639
Configuration 2									
j	$\Delta_j(3,4)$	$\gamma_j(4)$	$\Delta_j(4,5)$	$\gamma_j(5)$	$\Delta_j(5,6)$	$\gamma_j(6)$	$\Delta_j(6,7)$	$\gamma_j(7)$	$\Delta_j(7,8)$
1	0.04693	1.38	0.01801	1.43	0.00669	1.44	0.00247	1.44	0.00091
2	0.04945	1.38	0.01905	1.43	0.00707	1.44	0.00260	1.44	0.00096
3	0.13891	1.45	0.05087	1.43	0.01883	1.43	0.00698	1.43	0.00259
4	0.14607	1.43	0.05407	1.43	0.02009	1.43	0.00747	1.43	0.00277
5	0.51118	1.85	0.14144	1.78	0.04130	1.71	0.01259	1.66	0.00399
6	0.51284	1.83	0.14405	1.76	0.04240	1.70	0.01301	1.65	0.00415
7	0.95116	1.98	0.24167	1.83	0.06792	1.75	0.02017	1.69	0.00625
8	0.94685	1.97	0.24149	1.83	0.06787	1.75	0.02013	1.69	0.00623
9	1.00313	1.92	0.26573	1.79	0.07687	1.71	0.02346	1.65	0.00746
10	1.01493	1.92	0.26847	1.79	0.07761	1.71	0.02366	1.65	0.00752

Table 6.7: h-convergence of the eigenvalues computed with Elements LOhex.

Eigenvalue #	level 2	level 3	level 4	level 5	level 6
1	14.004191	14.000267	14.000017	14.000001	14.000000
2	14.004191	14.000267	14.000017	14.000001	14.000000
3	15.786221	15.779749	15.779334	15.779308	15.779306
4	15.786221	15.779749	15.779334	15.779308	15.779306
5	28.381166	28.347985	28.345822	28.345686	28.345677
6	28.381166	28.347985	28.345822	28.345686	28.345677
7	30.163195	30.127466	30.125139	30.124992	30.124983
8	30.163195	30.127466	30.125139	30.124992	30.124983
9	41.024619	40.919524	40.912522	40.912077	40.912049
10	41.024619	40.919524	40.912522	40.912077	40.912049

Table 6.8: The eigenvalues computed with Elements HOhex for Configuration 0.

j	$\Delta_j(2,3)$	$\gamma_j(3)$	$\Delta_j(3,4)$	$\gamma_j(4)$	$\Delta_j(4,5)$	$\gamma_j(5)$	$\Delta_j(5,6)$
1	0.003924	3.97	0.000251	3.99	0.000016	4.00	0.000001
2	0.003924	3.97	0.000251	3.99	0.000016	4.00	0.000001
3	0.006472	3.96	0.000415	3.99	0.000026	4.00	0.000002
4	0.006472	3.96	0.000415	3.99	0.000026	4.00	0.000002
5	0.033181	3.94	0.002163	3.98	0.000137	4.00	0.000009
6	0.033181	3.94	0.002163	3.98	0.000137	4.00	0.000009
7	0.035729	3.94	0.002327	3.98	0.000147	4.00	0.000009
8	0.035729	3.94	0.002327	3.98	0.000147	4.00	0.000009
9	0.105095	3.91	0.007002	3.98	0.000445	3.99	0.000028
10	0.105095	3.91	0.007002	3.98	0.000445	3.99	0.000028

Table 6.9: h-convergence of the eigenvalues computed with Elements HOhex for Configuration 0.

Eigenvalue #	level 3	level 4	level 5	level 6
1	3.80281	3.79506	3.79223	3.79119
2	4.07274	4.06450	4.06153	4.06044
3	4.83375	4.81193	4.80386	4.80087
4	5.19253	5.16903	5.16036	5.15716
5	9.52243	9.49485	9.48629	9.48325
6	9.79148	9.76188	9.75278	9.74956
7	10.98751	10.94893	10.93688	10.93256
8	11.12740	11.08827	11.07625	11.07196
9	12.48320	12.42976	12.41362	12.40789
10	12.63651	12.58233	12.56607	12.56030

Table 6.10: The eigenvalues computed with Elements HOhex for Configuration 2.

j	$\Delta_j(3,4)$	$\gamma_j(4)$	$\Delta_j(4,5)$	$\gamma_j(5)$	$\Delta_j(5,6)$
1	0.00775	1.45	0.00283	1.44	0.00104
2	0.00824	1.47	0.00297	1.45	0.00109
3	0.02181	1.43	0.00808	1.43	0.00299
4	0.02350	1.44	0.00867	1.43	0.00321
5	0.02758	1.69	0.00856	1.50	0.00304
6	0.02960	1.70	0.00910	1.50	0.00322
7	0.03858	1.68	0.01206	1.48	0.00432
8	0.03913	1.70	0.01202	1.49	0.00429
9	0.05344	1.73	0.01614	1.49	0.00573
10	0.05418	1.74	0.01627	1.50	0.00577

Table 6.11: h-convergence of the eigenvalues computed with Elements HOhex for Configuration 2.

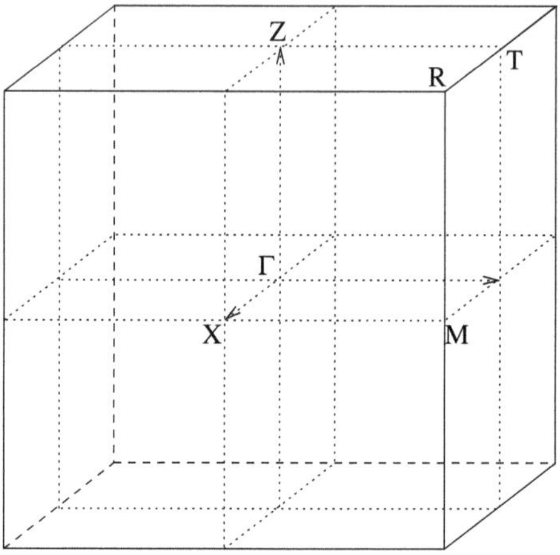

Figure 6.4: The Brillouin zone and the points of high symmetry.

6.5 Band structures

Our Maxwell code is able to compute the eigenvalues for a given material distribution and **k** point. Varying **k** along the boundary of the Brillouin zone one gets the band structure. The computed band structures can be used to proof existence of a spectral band gap and hence they may be of interest for physics and electronic engineering. A band structure with a band gap can also be used as a starting point for an optimization problem which e.g. try to expand the gap further (see e.g. [10]).

For the band structure computations we use Elements LOhex, the problem is solved on the refinement level 5 (the mesh is $32 \times 32 \times 32$). **k** changes uniformly along the lines between the points of high symmetry in the Brillouin zone (see Figure 6.4), in every interval $[X, Y]$ there are 3 intermediate points. The first 10 eigenvalues are computed up to the precision $\epsilon_E = 10^{-2}$.

We start with Configuration 0. Although this distribution cannot have a band gap it may be interesting to see how the band structure for a uniform medium looks like. The structure is presented in Figure 6.5. One may distinguish only 2-3 independent lines, the bands touch each other and merge into groups.

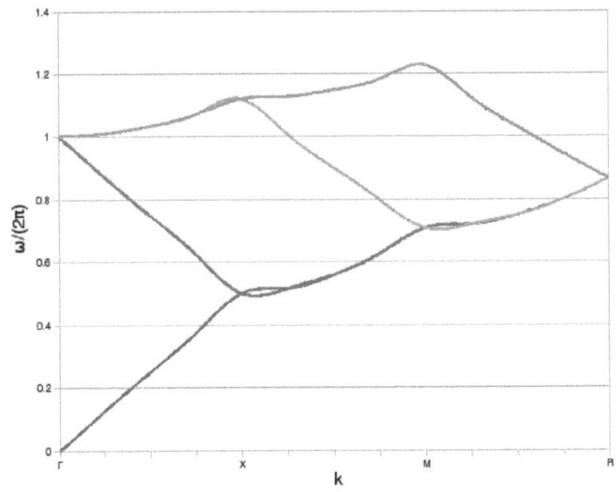

Figure 6.5: Band structure for Configuration 0, no band gap.

The band structure for Configuration 1 is shown in Figure 6.6. Due to the high symmetry of the distribution we go along just 4 **k** points. The band structure coincides with one presented in [12, Fig. 2], it has a moderate band gap.

The band structure for Configuration 2 is shown in Figure 6.7. Less symmetry in the distribution means that we need to go along more **k** points. This band structure exhibits a larger gap than one for Configuration 1. The structure is very similar with [12, Fig. 3]. They do not coincide perfectly because there is some difference in the distributions. In [12] the fundamental domain is $1 \times 1 \times \sqrt{2}$ what means that the block z size is $\sqrt{2}/4$, not $1/4$ as ours.

For the band structure computations we are interested in the eigenvalues only. The eigenfunctions are computed, but not used. Other applications in photonic crystals may require the eigenfunctions as well, e.g. the band gap optimization problem mentioned before need the eigenfunctions. A visualization of an eigenfunction is presented in Figure 6.8.

6.6 Computer-assisted proof for band gap

The method was explained in Section 1.3. We proved Theorem 1.7, which formulates a perturbation argument for the spectrum of $A_{\mathbf{k}}$. The following theorem express the same idea, but provides a better estimate.

Theorem 6.1. *(see [5])*
Let $B(\mathbf{k}, r) = \{\mathbf{k}' \in \mathbb{R}^3 \mid |\mathbf{k}' - \mathbf{k}| < r\}$ *and* $\varepsilon_{\min} = \min_{\mathbf{x}\in\Omega} \varepsilon(\mathbf{x})$. *Suppose that for the operator* $A_{\mathbf{k}}$ *and some* $l \in \mathbb{N}$ *there exists an interval* $[a, b]$ *s.t.*

1. $[a, b] \subset (\overline{\lambda}_{\mathbf{k},l}, \underline{\lambda}_{\mathbf{k},l+1})$ *for all* $\mathbf{k} \in \mathcal{K}$,

2. $K \subset \bigcup_{\mathbf{k}\in\mathcal{K}} B(\mathbf{k}, r_{\mathbf{k}})$, *where* $r_{\mathbf{k}}$ *holds*

$$\max\left\{1, \frac{1}{\varepsilon_{\min}} + r_{\mathbf{k}}\right\} \max\left\{\frac{\overline{\lambda}_{\mathbf{k},l} + 1}{a - \overline{\lambda}_{\mathbf{k},l}}, \frac{\underline{\lambda}_{\mathbf{k},l+1} + 1}{\underline{\lambda}_{\mathbf{k},l+1} - b}\right\} r_{\mathbf{k}} \leq 1,$$

then $[a, b]$ *is contained in a spectral gap, i.e.* $[a, b] \subset (\overline{\lambda}_{\mathbf{k},l}, \underline{\lambda}_{\mathbf{k},l+1})$ *for all* $\mathbf{k} \in K$.

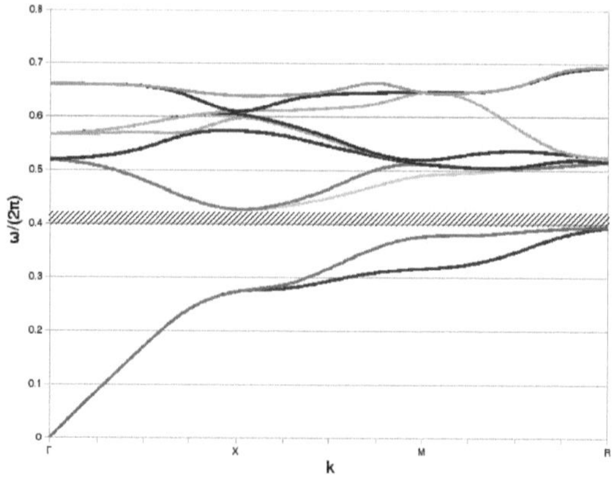

Figure 6.6: Band structure for Configuration 1, a band gap between the bands 2 and 3.

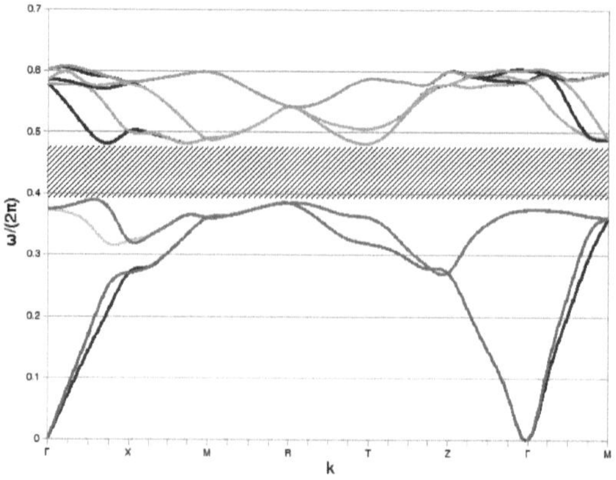

Figure 6.7: Band structure for Configuration 2, a band gap between the bands 4 and 5.

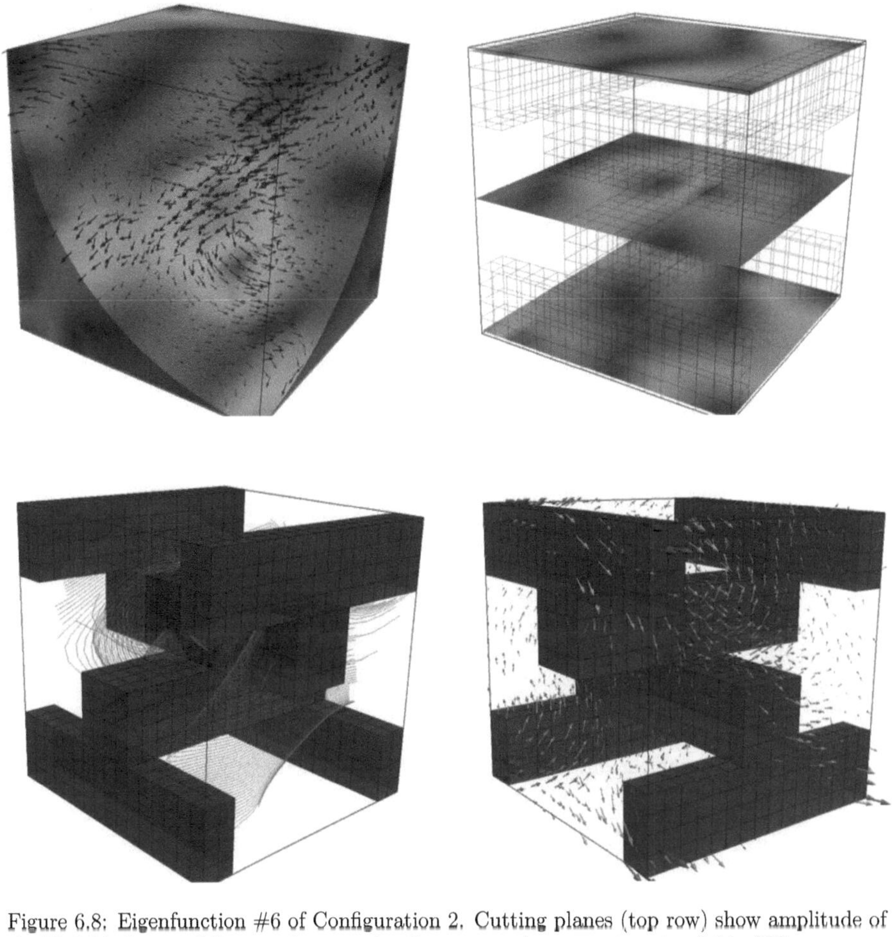

Figure 6.8: Eigenfunction #6 of Configuration 2. Cutting planes (top row) show amplitude of the **H** field, "stream"-lines and arrow-plot (bottom row) represent vector field of **H**.

In few words, the radius $r_{\mathbf{k}}$ of a safe perturbation of $\lambda_{\mathbf{k},j}$ is larger for $\underline{\lambda}_{\mathbf{k},l+1}, \overline{\lambda}_{\mathbf{k},l}$ which are far from the band gap, and smaller for those that are close to. Another point is that the radius is larger if we allow some more tolerance for the precise band gap boundaries $[a, b]$, the tolerance in its turn is limited by the band gap width. So the theorem gives a very natural conclusion.

Below we provide computational results which illustrate a practical application of the theorem. We consider Configuration 1. Since at the moment there are no $\overline{\lambda}_{\mathbf{k},l}$ and $\underline{\lambda}_{\mathbf{k},l+1}$ available for the 3D problem we replace them with $\lambda_{\mathbf{k},h,l}$ and $\lambda_{\mathbf{k},h,l+1}$ computed by our eigenvalue solver on level 4 with the accuracy $\epsilon_E = 10^{-4}$. The band gap is located between the 2-nd and 3-d bands. On this level $\max_{\mathbf{k}\in\mathcal{K}}\{\lambda_{\mathbf{k},h,2}\} = 6.1553$ and $\min_{\mathbf{k}\in\mathcal{K}}\{\lambda_{\mathbf{k},h,3}\} = 7.3043$, so the approximated band gap width is 1.149. According to Table 6.6 sum of the differences between the level 4 and 8 for the 2-nd and 3-rd eigenvalues is about 0.2495, so the approximated band gap width might become about 0.8995 on level 8, but it still can be used if we assume some higher tolerance. Let us define the almost maximal tolerance and put $a = 6.6432$, $b = 6.6913$. For practical purposes it would be a very narrow band gap, but it is sufficient since we only want to show the technology. Putting all the parameters in Theorem 6.1 one gets

$$r_{\mathbf{k}}^2 + r_{\mathbf{k}} - \frac{1}{M_{\text{dist}}} \leq 0, \qquad r_{\mathbf{k}} > 0,$$

where $M_{\text{dist}} = \max\left\{\frac{\lambda_{\mathbf{k},h,l}+1}{6.6432-\lambda_{\mathbf{k},h,l}}, \frac{\lambda_{\mathbf{k},h,l+1}+1}{\lambda_{\mathbf{k},h,l+1}-6.6913}\right\}$. It gives the solution

$$r_{\mathbf{k}} \leq \frac{1}{2}\left(\sqrt{1+\frac{4}{M_{\text{dist}}}} - 1\right). \tag{6.4}$$

Now we proceed with the computations. The irreducible Brillouin zone of Configuration 1 is a tetrahedron with the vertices $K = (\Gamma, X, M, R)$. From (6.4) one may derive the worst case radius and then define a regular mesh satisfying that radius. But this approach requires too many $\mathbf{k}$ points and therefore leads to too much extra computations. A better idea is an adaptive choice of $\mathbf{k}$ points. We suggest the following simple adaptive method.

In K we define a very fine regular mesh $\tilde{\mathcal{K}}$ with $dk = 0.01$ (satisfying the worst case radius), also set $\mathcal{K} = \varnothing$. Step by step we go along all $\mathbf{k} \in \tilde{\mathcal{K}}$. If $|\mathbf{k}-\mathbf{k}'| > r_{\mathbf{k}'}$ for all $\mathbf{k}' \in \mathcal{K}$, then compute the eigenvalues $\lambda_{\mathbf{k},h,j}$ and add $\mathbf{k}$ to $\mathcal{K}$, else go to the next $\mathbf{k}$. In such a way we build a suboptimal covering set $\mathcal{K}$ mentioned in Theorem 6.1. The coverage is shown at Figure 6.9. In our example $\mathcal{K}$ consists of 1523 points. The total computational time is about 12 hours on 8 processors. The minimal perturbation radius is $r_{\mathbf{k}} = 0.064$, the maximal one is $r_{\mathbf{k}} = 0.271$. In general, depending on $[a, b]$ the number of $\mathbf{k}$ points and computational time may vary significantly.

Remember that the computations described above is not the proof. For the actual proof one needs correct estimates $\overline{\lambda}_{\mathbf{k},l}$ and $\underline{\lambda}_{\mathbf{k},l+1}$.

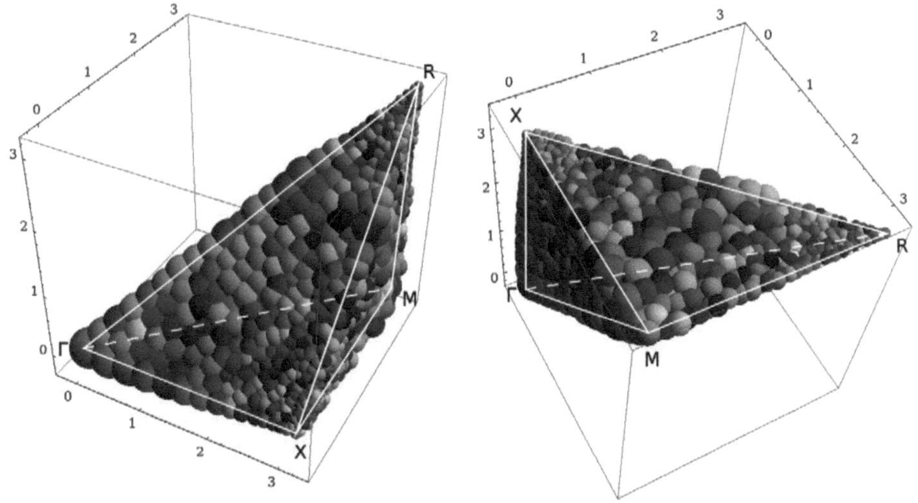

Figure 6.9: The irreducible Brillouin zone covered by the system of balls for the perturbation proof.

Bibliography

[1] E. Anderson, Z. Bai, C. Bischof, S. Blackford, J. Demmel, J. Dongarra, J. Du Croz, A. Greenbaum, S. Hammarling, A. McKenney, and D. Sorensen. *LAPACK user's guide.* SIAM, third edition, 1999. http://netlib.org/lapack/.

[2] D.N. Arnold, R.S. Falk, and R. Winther. Multigrid in H(div) and H(curl). *Numerische Mathematik*, 85:197–218, 2000.

[3] Z. Bai, J. Demmel, J. Dongarra, A. Ruhe, and H. van der Vorst, editors. *Templates for the solution of algebraic eigenvalue problems: a practical guide.* SIAM, 2000.

[4] R. Barrett, M. Berry, T. Chan, J. Demmel, J. Donato, J. Dongarra, V. Eijkhout, R. Pozo, C. Romine, and H. van der Vorst. *Templates for the solution of linear systems: building blocks for iterative methods.* SIAM, 1994.

[5] Henning Behnke, Malcolm B. Brown, Vu Hoang, and Michael Plum. A computer-assisted band-gap proof for 3D photonic crystals. To be published.

[6] Daniele Boffi. Discrete compactness and Fortin operator for edge elements. *Numerische Mathematik*, 87:229–246, 2000.

[7] Daniele Boffi, Franco Brezzi, and Lucia Gastaldi. On the convergence of eigenvalues for mixed formulations. *Ann. Sc. Norm. Sup. Pisa*, 25:131–154, 1997.

[8] Daniele Boffi, Matteo Conforti, and Lucia Gastaldi. Modified edge finite elements for photonic crystals. *Numerische Mathematik*, 105(2):249–266, 2006.

[9] J. Bramble, D. Kwak, and J. Pasciak. Uniform convergence of multigrid V-cycle iterations for indefinite and nonsymmetric problems. *SIAM Journal on Numerical Analysis*, 31:1746–1763, 1994.

[10] Steven J. Cox and David C. Dobson. Maximizing band gaps in two-dimensional photonic crystals. *SIAM Journal on Applied Mathematics*, 59(6):2108–2120, 1999.

[11] L. Demkowicz, P. Monk, L. Vardapetyan, and W. Rachowicz. De Rham diagram for hp finite element spaces. *Computers and Mathematics with Aplications*, 39(7–8):29–38, 2000.

[12] David C. Dobson, Jayadeep Gopalakrishnan, and Joseph E. Pasciak. An efficient method for band structure calculations in 3D photonic crystals. *Journal of Computational Physics*, 161(2):668–679, 2000.

[13] David C. Dobson and Joseph E. Pasciak. Analysis of an algorithm for computing electromagnetic Bloch modes using Nedelec spaces. *Computational Methods in Applied Mathematics*, 1(2):138–153, 2001.

[14] Wolfgang Hackbush. *Multigrid methods and applications.* Springer-Verlag, 1985.

[15] Wolfgang Hackbush. *Iterative solution of large sparse system of equations.* Springer-Verlag, 1994.

[16] Ralf Hiptmair. Multigrid method for Maxwell's equations. *SIAM Journal on Numerical Analysis*, 36(1):204–225, 1998.

[17] Ralf Hiptmair. Finite elements in computational electromagnetism. *Acta Numerica*, 11:237–339, 2002.

[18] Ralf Hiptmair and Klaus Neymeyr. Multilevel method for mixed eigenproblems. *SIAM Journal on Scientific Computing*, 23(6):2141–2164, 2002.

[19] Vu Hoang, Michael Plum, and Christian Wieners. A computer-assisted proof for photonic band gaps. *Z. Angew. Math. Phys.*, 60:1–18, 2009.

[20] John David Jackson. *Classical electrodynamics.* John Wiley & Sons, 1975.

[21] John D. Joannopoulos, Steven G. Johnson, Joshua N. Winn, and Robert D. Meade. *Photonic crystals: molding the flow of light.* Princeton University Press, 2008.

[22] Charles Kittel. *Solid state physics.* John Wiley & Sons, 1986.

[23] A. V. Knyazev, M. E. Argentati, I. Lashuk, and E. E. Ovtchinnikov. Block locally optimal preconditioned eigenvalue xolvers (BLOPEX) in Hypre and PETSc. *SIAM Journal on Scientific Computing*, 29(5):2224–2239, 2007.

[24] Andrew V. Knyazev. Toward the optimal preconditioned eigensolver: locally optimal block preconditioned conjugate gradient method. *SIAM Journal on Scientific Computing*, 23(2):517–541, 2001.

[25] Andrew V. Knyazev and Klaus Neymeyr. A geometric theory for preconditioned inverse iteration. III: A short and sharp convergence estimate for generalized eigenvalue problems. *Linear Algebra and its Applications*, 358(1):95–114, 2003.

[26] Peter Kuchment. *Floquet theory for partial differential equations*, volume 60 of *Operator theory advances and applications.* Birkhäuser Verlag, 1993.

[27] Peter Kuchment. The mathematics of photonic crystals. *Mathematical modeling in optical science*, pages 207–272, 2001.

[28] Peter Monk. *Finite element methods for Maxwell's equations.* Oxford University Press, 2003.

[29] Jean-Claude Nédélec. Mixed finite elements in $\mathbb{R}^3$. *Numerische Mathematik*, 35(3):315–341, 1980.

[30] Jean-Claude Nédélec. A new family of mixed finite elements in $\mathbb{R}^3$. *Numerische Mathematik*, 50(1):57–81, 1986.

[31] Klaus Neymeyr. A geometric theory for preconditioned inverse iteration I: Extrema of the Rayleigh quotient. *Linear Algebra and its Applications*, 322:61–85, 2001.

[32] Klaus Neymeyr. A geometric theory for preconditioned inverse iteration II: Convergence estimates. *Linear Algebra and its Applications*, 322:87–104, 2001.

[33] Farouk Odeh and Joseph B. Keller. Partial differential equations with periodic coefficients and bloch waves in crystals. *Journal of Mathematical Physics*, 5(11):1499–1504, 1964.

[34] Beresford N. Parlett. *The symmetric eigenvalue problem*, volume 20 of *Classics in applied mathematics*. SIAM, 1998.

[35] J. SchÂĺoberl. *Robust Multigrid Methods for Parameter Dependent Problems*. PhD thesis, Johannes Kepler Universität Linz, 1999.

[36] G. Szegö. *Orthogonal polynomials*, volume 23 of *Colloquium Publications*. American Mathematical Society, third edition, 1974.

[37] Rajeev Thakur, Rolf Rabenseifnery, and William Gropp. Optimization of collective communication operations in MPICH. *Int. J. High Perform. Comput. Appl.*, 19(1):49–66, 2005.

[38] A. Toselli and O. Widlund. *Domain decomposition methods - algorithms and theory*, volume 34 of *Computational Mathematics*. Springer-Verlag, 2005.

[39] Christian Wieners. A geometric data structure for parallel finite elements and the application to multigrid methods with block smoothing. To be published.

[40] Jinchao Xu. Iterative methods by space decomposition and subspace correction. *SIAM Review*, 34(4):581–613, 1992.

[41] Sabine Zaglmayr. *High order finite element methods for electromagnetic field computation*. PhD thesis, Johannes Kepler Universität Linz, 2006.

MIX
Papier aus verantwortungsvollen Quellen
Paper from responsible sources
FSC® C105338

Printed by Books on Demand GmbH, Norderstedt / Germany